THE GREEN MACHINE

"In these twentieth-century days, a careful municipality has studded the down with rustic seats and has shut its dangers out with railings, has cut a winding carriage-drive round the curves of the cove down to the shore, and has planted sausage-laurels at intervals in clearings made for that aesthetic purpose. When last I saw the place, thus smartened and secured, with its hair in curl-papers and its feet in patent-leathers, I turned from it in anger and disgust, and could almost have wept. I suppose that to those who knew it in no other guise, it may still have beauty. No parish councils, beneficent and shrewd, can obscure the lustre of the waters or compress the vastness of the sky. But what man could do to make wild beauty ineffectual, tame and empty, has amply been performed at Oddicombe."

EDMUND GOSSE
Father and Son, 1907

I dedicate this book to the wild beauty of unmolested nature and to all those who seek to ensure its survival.

THE GREEN MACHINE

Ecology and the Balance of Nature

WALLACE ARTHUR

Basil Blackwell

British Library Cataloguing in Publication Data
A CIP catalogue record for this book is available from the British Library.

Library of Congress Cataloging in Publication Data
Arthur, Wallace.
The green machine: ecology and the balance of nature / Wallace
Arthur.
        p.   cm.
Includes bibliographical references.
ISBN 0–631–17853–8 (pbk)
        1. Biotic communities.   2. Ecology.   3. Man – Influence on nature.
I. Title.
QH541.A725   1990
574.5 – dc20
                                                                              90–1568
                                                                              CIP
Typeset in 11 on 13pt Sabon
by Wearside Tradespools, Fulwell, Sunderland
Printed in Great Britain by
TJ Press Ltd., Padstow

Contents

PREFACE

Most prefaces are boring and redundant, and would have been better left out. This one's different, because it does something useful: it gives you an insight into the structure of the book and the kinds of reading strategy that could be adopted by different kinds of reader.

The book is in four parts, though they're not explicitly labelled as such. The first (Chapters 1 to 5) sets the scene by looking at why we should study ecological systems, and painting a very broad picture of the form those systems take. The second (Chapters 6 to 11) provides an in-depth look at the way animals and plants interact with each other, in natural communities, to produce what we call the balance of nature. The third (Chapters 12 to 14) adds an evolutionary dimension by looking back at the history of the ecosphere over timespans ranging from a few generations to four billion years. The fourth (Chapters 15 and 16) considers problems that human activities pose to the fabric of ecosystems, including extinctions, 'agriculturalization' and the threat of wide-scale radioactivity. Chapter 17 is a sort of 'grand finale'.

So how should you read it? Well, like most books, it's designed primarily to be read from cover to cover, but that's not the only sensible strategy. There are at least three other approaches. If you're mainly interested in topical green issues and don't want to get too technical, skip the second part and dip into it later if you feel like it. If you're of the opposite disposition, and want to find out how nature works, when unhindered by human activities, skip the fourth part, and use it just for reference when you're later reading a newspaper article and it doesn't tell you what 'global warming' (or whatever) actually is. And if you're interested both in unfettered nature *and* in environmental issues arising out of human activities, but you're already familiar with ecological fundamentals, skip the first part (after Chapter 1) and pick up the thread at Chapter 6.

Well, that's it. No acknowledgements (they're elsewhere), no repetition of points I'm going to make later, and no excuses for the things I've left out. Just a warning that if you read on you'll be taken on a journey into space.

Wallace Arthur

Acknowledgements

Like most books, this one started as an idea which became transmuted into a scribble. From scribbled idea to print is a long and difficult road. I have been greatly assisted in reaching the end of it by Mark Allin, whose enthusiasm right from the beginning helped enormously, and Shelagh Tunney, whose word-processing skills prevailed over the succession of difficulties that a randomly-evolving computer system conspired to throw at her. Others, too numerous to name, have put up with my incessant questions about how best to present ecology to the general reader.

While most of the figures that appear here are original, several are not, and I am indebted to the following for permission to reproduce (with modification in some cases) various illustrations. Figure 1 is reproduced by permission from J. H. Brown and A. C. Gibson, *Biogeography* (St Louis, 1983), the C. V. Mosby Co.; Figure 10 (top) is from P. Williamson, R. A. D. Cameron and M. A. Carter, 'Population density affecting adult shell size of snail *Cepaea nemoralis*', reprinted by permission from *Nature*, vol. 263, pp. 496–7, copyright © 1976 Macmillan Magazines Ltd; Figure 10 (bottom) is from J. P. Dempster, 'The ecology of the cinnabar moth', *Advances in Ecological Research*, 12 (1982), by courtesy of Academic Press; Figures 14, 36 and 37 (bottom) are from *Ecology: The Experimental Analysis of Distribution and Abundance*, 3rd edn, by Charles J. Krebs, copyright © 1985 by Harper & Row, Publishers, Inc., reprinted by permission of Harper & Row, Publishers, Inc.; Figures 15 and 22 are from W. Arthur, *The Niche in Competition and Evolution* (1987), reprinted by permission of John Wiley & Sons Ltd; Figure 18 is from M. Hassell, *The Dynamics of Competition and Predation* (Arnold, 1976); Figure 23 (bottom) is from M. B. Usher and M. Williamson (eds), *Ecological Stability* (Chapman & Hall, 1974); Figures 25, 31, 37 (top) and 39 are from *Evolutionary Ecology* by Eric R. Pianka, copyright © 1988 by Harper & Row, Publishers, Inc., reprinted by permission of Harper & Row, Publishers, Inc.; Figures 26 and 41 are from *Fundamentals of Ecology*, Third Edition, by E. P. Odum, copyright © 1971 by Saunders College Publishing, a division of Holt, Rinehart and Winston, Inc., reprinted by permission of the publisher.

I

WONDER VERSUS ANXIETY

Imagine that you are an astronaut with a spacecraft more typical of science fiction than of current reality – one that can take you on journeys of your choice wherever you wish to go, unimpeded by the temporary distance limitations imposed by Einsteinian physics. Using this craft you embark on a journey of discovery, first visiting neighbouring planets in our own solar system (Mars, Jupiter and so on), then venturing further afield and haphazardly wandering the galaxies visiting planets here and there to sample the flavour of the universe. What would this flavour turn out to be?

Answering this question is inevitably guesswork since the journey has yet to be undertaken, but the best guess is that the central strand in the flavour of the universe is 'staticness', to coin a more forceful, albeit less grammatical, version of 'stasis'. Now at first this must seem a weird proposition. After all, the various heavenly bodies are in a state of perpetual motion to such an extent that nothing can be considered as a fixed reference point in relation to which the motion of other objects can be measured. If all is motion, how can the prevailing flavour of the universe be staticness? In order to see the sense of this apparent contradiction we need to view the universe not as an astronomer but as an ecologist. We need to see it through the eyes of some future Ecological Officer on board an intergalactic mission, or through our own eyes on the hypothetical journey with which we started.

The ecologist does not sit back in some cosmic armchair, viewing the universe from afar and thus perceiving an ever-swirling kaleidoscope of objects of varying brightness, each treading their ordained circuits through the void. Rather, he takes the 'hands-on' approach, and perceives each piece of the universe that he visits face-to-face much as you or I would perceive a chunk of African jungle into which we were dropped by parachute. Through the ecologist's eyes, then, the flavour of the universe is not the grand overall scheme of things but rather the flavour of the typical planetary environment. Just as contemporary terrestrial ecologists (meaning non-cosmic rather than non-aquatic ones) build up a picture of, for example, the typical 'coniferous forest environment' by a process of sampling small, workable bits of it here and there, so their space-tramping colleagues build up a picture of the typical planetary environment by a similar sort of random sampling procedure.

The vision of staticness throughout the universe now begins to make more sense, and my job here is to make you almost *feel* it. Imagine once again that you have embarked on an exploratory universal voyage, aided by your 3001-model spacecraft. What would your typical experience be on leaping out of the just-landed craft and looking around you? Our best model for such an experience is the Martian environment, as perceived by that robotic ecologist, *Viking*. This experience suggests that you'd be staring out across a barren landscape of rock and dust. However closely you looked, nothing would move except, perhaps, for the dust disturbed by your space-boots. There would be no trees blowing in the wind, no birds singing, no insects casually buzzing past your face. You would be experiencing an environment with a degree of staticness unmatched by anything on earth.

The sceptic might argue, of course, that this is a superficial view. The 'eyeballing' of a landscape will fail to detect all kinds of hidden activities. There could, for example, be intense microbial activity taking place all around you in the dust, and you'd be none the wiser. Anticipating such a sceptic, you would doubtless have packed a microscope and other means of monitoring that are more powerful than the human eye; but the picture you'd get on using your microscope would be the same as before, merely on a different scale. A typical field of vision in your microscope would reveal a lot of static dust – nothing happening, no small creatures moving, no spark of life.

We shouldn't, perhaps, take 'static' too literally. Mars has an atmosphere of sorts, and can thus experience wind. Also, it will undergo geological activities such as sedimentation, weathering and so on. But such processes take place on such an immense timescale that they scarcely affect our perception of a barren, static world.

As any ecologist knows, you can't get too far in your understanding of an ecological system using a sampling procedure if your sample size is 1. In fact, with some qualifications that I won't bore you with, the bigger the sample size the more reliable is your information. Consequently, if you set out to get a feel for the flavour of the universe in the sense of what the typical planetary environment is like, you wouldn't want to stop at Mars. You might well start there – indeed, if your journey was 'outward' from earth, Mars would be the logical first sample. And an outward journey (in solar system terms) makes considerably more sense than an 'inward' one, which would give you a sample size of 2 (Venus, Mercury) before you were obliterated by nuclear fusion on reaching the sun.

A sensible sampling programme (we'll see the criteria for this later) might take you through the outer planets of our system, then on through a series of randomly chosen systems elsewhere in our Milky Way galaxy, and finally through yet more random choices even further afield, before returning home suffering severely from space-lag and with enough samples of material to keep your earth-bound laboratory busy for years.

How typical would the Martian experience be, both at the jumping-out-of-the-spacecraft level and at the microscopical or biochemical analysis level? My guess is that it would be very typical indeed, to the extent that almost every planet you visited would present you with the same overall picture – rocks and dust. Of course, differences in detail would abound. Colours would vary from planet to planet, as would the temperature. Landscapes might be uninterrupted flat plains stretching boringly to the horizon, or they might be strikingly contoured by a myriad of mountains and craters of different sizes and shapes. But all – or almost all – would be static and lifeless.

One little nicety that should perhaps be emphasized is that some planets couldn't actually be visited in such a way that you could achieve your jumping-out-of-the-spacecraft perception. There is a very good reason for this: some planets are purely gaseous, with the consequence that anyone attempting to jump on to them would be in for a bit of a shock and a premature termination of their sampling programme. However, such planets are even less likely to provide a home for the frenzied activities of living creatures – whether visible or microscopical – than solid ones like Mars. So the flavour of a static, lifeless universe still prevails.

Now it has to be said that the ecologist's view of the universe, as I have described it, is an exceedingly biased one. It is centred, as we have seen, on the 'planetary environment', which means the interface between the substrate and the atmosphere. Such a view leaves out of consideration what happens in planetary cores, in upper atmospheres, in the suns that power the solar systems in which planets are embedded, and in interstellar space. But there's a good reason for this bias. Ecologists, like other biologists and naturalists, are interested in living creatures; and their scrutiny of the universe – both current/imaginary and future/real – is aimed at detecting potential ecospheres equivalent to our terrestrial one (sometimes alternatively called the biosphere) which extends from a few metres underground up into the lower atmosphere. Of course, it's *possible* that some planets will turn out to have living creatures near their core or somewhere else where we wouldn't expect them, rather than around their surface. But such a possibility seems sufficiently remote that we can disregard it until some pressing reason suggests otherwise.

From Wonder to Questions

By now you may justifiably have come to the conclusion that the Green Machine is a spacecraft used in the fruitless search for potential ecospheres on what turn out to be lifeless planets, but thankfully it is not. If it was, then a book about its adventures would hold about as much interest for the ecologically-minded reader as a lump of lead. My galactic opening is simply a means to an end, and the end is a contrast which, from our parochial, earth-based viewpoint, can easily be missed if steps are not taken at the outset to expand our range of vision. The contrast that I wish to emphasize is this. On returning from your multi-stop tour of the universe, your mind would be full of pictures of static, barren, lifeless landscapes. You might have visited a thousand planets, yet all would be the same in this respect. Now suppose that on re-entering the earth's atmosphere after several years' absence your navigation systems fail and you end up landing safe but lost in a small clearing in the rainforests of Borneo.

Imagine your feelings on doing your jumping-out-of-the-spacecraft act. The perpetual motion of life would hit you like a hammer. Within seconds of climbing out of your now-redundant spacesuit your flesh would be attacked by a range of unwelcome insect visitors. As you walked towards the jungle, drawn by its magnetism from your small clearing, there would be a rustling of leaves as diverse small creatures ran from the threat of the unknown. Close inspection of the vegetation would reveal a host of crawling things all going unconsciously about their business. If you were lucky, you might even spot a lone orang-utan scratching its head and viewing you philosophically from its favourite tree. Microscopy and biochemical analysis would compound this picture of perpetual motion by revealing, even within the superficially static plants, cells buzzing with chemical activity operating on an infinitely faster timescale than the sluggish chemistry of the Martian rocks – a living chemistry tuned and re-tuned by competition for survival over the aeons. Finally, after years of nothingness, you would be overwhelmed by the intensity of biological motion at all levels. It would be an intoxicating experience indeed.

Out of this contrast jumps one of the emotions that I used in the title of this chapter: wonder. Wonder at the extraordinary motion, complexity and interaction of the ecosphere. A confusion of birth and death, sex and violence, organic construction and decay. In a five-minute period corresponding, perhaps, to the period that you stood transfixed by the nothingness of the Martian landscape several years ago, thousands of creatures will have died and been born even within the tiny area of jungle that you can see around you. In contrast to the rest of the universe we have a seething, frenzied, perpetual activity, some of it visible to the naked eye,

5

much of it not, since the vast majority of creatures on earth are microscopic.

Of course, I've deliberately sought to maximize the contrast by crash-landing you in the Bornean jungle instead of in the Arctic Ocean or the Sahara Desert. Yet even these least promising parts of the ecosphere are populated with a variety of interacting life forms, as any serious attempt to study them readily reveals. Contrasts between different parts of the ecosphere may at first appear dramatic – especially when extreme environments are compared – but they fade into insignificance when *any* terrestrial environment is compared with a Martian, or other non-terrestrial, equivalent. It is the whole ecosphere that is dynamic with life, not just a few selected parts of it.

Perception of the ecosphere as a scene of almost incredible living activity in contrast to the staticness that abounds elsewhere immediately leads to a whole host of questions. How does it work? Where did it come from? What causes the different degrees of living activity in different places? How stable is it all? One could go on almost endlessly, especially if we make the questions a little more specific. The contrast I've tried to emphasize gives rise to wonder, and the wonder to questions. Attempting to answer these questions is the core of what can be called 'ungreen ecology', which is the phrase I use to describe the ecology of wonder as opposed to the ecology of anxiety.

The reason for this choice of phrase is probably obvious. Over the last twenty years, the ecology of anxiety – about the fragility of the ecosphere and man's cavalier attitude to it – has grown and grown. It is now a major political force in some of the most supposedly 'advanced' countries where environmental damage is also most 'advanced'. And everywhere it goes, the ecology of anxiety uses the 'Green' label. In the English-speaking world, someone talking about 'greens' is now as likely to be speaking of an ecopolitical movement as they are to be referring to vegetables. In Germany, we have die Grünen, and so on. The ecology/green connection is now so firmly established in the popular press that many non-ecologists are tempted to equate the two.

Such an equation, as we've seen, is false. Ecology is more than just the scientific basis for green politics: that is only part of it.

Ecology is also the attempt to *understand* the extraordinary 'specialness' of the ecosphere and the unique processes going on within it. This ecology – the ecology of wonder, or 'ungreen' ecology – is the main subject of this book. I will indeed have more to say about green ecology and ecopolitics, in Chapters 15 and 16 which are specifically devoted to green issues. But there is no need for yet another complete book on the ecology of anxiety – there are enough of those already.

I do not, it must be stressed, wish to set up some false antagonism between the two 'halves' of ecology or to give the impression that to do ungreen ecology is to be anti-green. Far from it. The kinds of issues that are central to green politics are vital to the survival and welfare of humanity as a whole. Who would argue that the environmental damage caused by large-scale nuclear accidents, acid rain, and the profligate dousing of vast areas of crops with extremely toxic chemicals is inconsequential? No-one with any degree of sanity and long-term vision, to be sure. I must admit to occasional annoyance about the *way* in which green activists go about their business, such as claiming a link between acid rain and forest destruction before there is any evidence to prove that it is the acid, and not something else, that is the destructive agent. But this isn't an argument against green politics in particular. After all, red and blue politicians are equally good if not better practitioners of the art of the unsubstantiated claim. So my annoyance arises simply out of the traditional enmity that someone brought up on scientific methods feels towards those who practice the more nebulous and transient business of politics.

Two Caveats

Our initial foray into space and subsequent landing in the jungle might have given rise to two misconceptions which are, in a sense, opposites. These are first, that the processes unique to the ecosphere are exclusively ecological; and second, that processes going on outside the (terrestrial) ecosphere are exclusively unecological. Let's have a look at just why these are indeed misconceptions.

With regard to our own ecosphere, a clue to the problem can be found in its alternative title – biosphere. The uniqueness of the processes going on within this 'sphere' (which it isn't) is certainly *biological* but it is not exclusively ecological. That is, to some extent biologists of other disciplines may be motivated by a sense of wonder at the uniqueness of biospheric processes occurring at levels other than the ecological – for example the biochemical level. However, ecology – broadly defined – ranges widely over a very large proportion of the biological sciences. There are established disciplines of eco-genetics, eco-physiology, behavioural ecology, evolutionary ecology, ecological biochemistry and so on. In a way, ecology is a frame of mind rather than a narrow discipline, and its practitioners may often be found using techniques that in other contexts would be described as biochemical, physiological or whatever.

Another thing about ecology that gives it prior claim over the other biological sciences to the sense of wonder that arises out of our interplanetary comparisons is that in ecological thinking context and location are of the essence. It is often said of biochemists working on enzymes extracted from rat liver mitochondria that it doesn't matter where the rats concerned came from, or whether, indeed, they used a pig's intestine as the source of their enzymes instead. As with many interdisciplinary jibes this one may not be entirely fair, but it serves to make a point. Finally, ecologists have primacy of possession of the interplanetary contrast by virtue of the fact that it is really an *ecological* comparison. Biochemists interested in comparative work can compare one enzyme with another, geneticists one gene with another, anatomists one kind of animal with another, and so on. It is the ecologist, when he searches around for a comparison involving ecospheres, who hits the full force of the interplanetary contrast.

We turn now to the second, 'opposite' misconception that I may have unintentionally encouraged – that everything outside the terrestrial ecosphere is unecological or, to put it another way, that the earth is the only single planet in the universe to possess an ecosphere and the life-forms which comprise it. While we are now almost certain that the earth is the only life-supporting planet in our solar system, that knowledge is matched by almost total ignorance of whether living organisms exist further afield. My guess is that there are other ecospheres, simply because in the vastness of the universe with its billions of planets it seems unlikely that life is restricted to just one of them. If this guess turns out to be true, then there will be a fascinating study of extra-terrestrial ecology to be undertaken in the future. Moreover, such a study will not only reveal specific details of the newly-discovered system but will also cast light on the degree of generality of ecology as a science, because we'll be able to test whether ecological principles that are true here are true there also, in the same way that the principles of physics seem to be applicable to physical matter however distant. I'll make a few speculations on this topic in the final chapter.

The acknowledgement that there may be a *few* extra-terrestrial ecospheres in no way lessens the force of the interplanetary contrast that I've tried so hard to emphasize – it simply realigns it so that the contrast is between the vast majority of lifeless planets and the few that do act as host to life, of whatever kind. And I still suspect, even after admitting the possibility of other ecospheres, that our 1000-planet sample would contain no other cases of life than here on earth. Again, we are in the world of guesswork, but while we're at it, my guess, for what it's worth, would be that the proportion of planets that possess ecospheres is *much* less than one in a thousand.

Ecology versus Natural History

If 'ungreen ecology' is not ecopolitics and is centred on an intense feeling of wonder about the uniqueness of the natural world that we see around us, does that mean that it's simply a new and trendy name for natural history? The answer to this question is a categorical 'no', despite the fact that ungreen ecologists and natural historians study the same thing (interactions between organisms and their environment) and for the same reason (the sense of wonder). In what way, then, are the two pursuits different from each other?

The answer to this question is best approached by examining a comment made by the American ecologist Robert J. Taylor, as follows: "General theory in ecology is the theory of a process reduced to its essence, stripped of the ornamentation of natural history."[1] The point is really this: natural history maximizes detail, ecology minimizes it. Someone interested in the natural history of blackbirds, cabbage white butterflies, or garden snails will study the beasts concerned, observing in detail what they look like and what they do, and will often carefully record details which will differ depending on the beast, but for the bird would include nest sites, food items, courtship and territorial behaviour. In natural history the details are an important, indeed essential, part of the endeavour, and they are acknowledged to differ from one species to another.

Ecologists, on the other hand, are interested in focusing in on aspects of the interaction between organism and environment that do *not* differ significantly between species. By 'ecologists' here (and later, when the word is used unqualified) I am referring to those ecologists working in the core of ungreen ecology – 'pure ecology', if you like – where the central concern is to understand the principles behind the structure and operation of the living world. An example of a general process of interest to such an ecologist is the S-shaped pattern of population growth often found when a species invades a new environment. The population grows slowly at first due to the very small number of colonists, then accelerates as the number of animals (or plants) breeding in the new environment increases, only to slow up again and finally stop growing as it approaches the maximum size that the environment can sustain.

From the ecologist's viewpoint, the fascinating thing is that there is a fundamental growth *pattern* which is repeated over and over again when different organisms invade different environments. In contrast, the natural historian might be more interested in the species-specific features that are superimposed on this pattern.

As with the green-versus-ungreen distinction, the naturalist-versus-ecologist one should not be taken to imply animosity between the two traditions concerned. Indeed, to be a good ecologist it helps considerably if you are a good natural historian too. The compilation of general principles may well involve the throwing away of unnecessary detail, but it's a *selective* discarding rather than a wholesale one that is needed – and the more details you know, the better a feel you are likely to have for which should be discarded in the making of a general principle and which retained.

And so to the Fray

I hope that by this stage I have provided sufficient background for our study of the Green Machine, and have imparted something of the sense of wonder at the unique or near-unique processes of the ecosphere which is close to the heart of all ecologists. The rest of the book is largely devoted to the attempts that have been, and are being, made to understand these processes. As we'll see, these attempts involve all sorts of skills: the skill of asking the right question in the first place; the skill of making detailed observations on highly complex systems; the skill of knowing when and how to profitably use simpler systems that permit experimental analysis as aids to understanding more complex ones; and, as we've seen, the skill of knowing which of the details of natural history to retain and which to discard in the search for general ecological principles.

Of course, we have a long way to go and ecology as a science is still in its infancy. But this leads to excitement as well as frustration. A science in which all the general principles have been established and tested long ago is at risk of becoming dry and boring. A science in which some principles are established, others tentative, and still others as yet unguessed-at may cause those practising it or reading about it to occasionally thump their desk in consternation – but boring it is certainly not.

2

THE AWFUL COMPLEXITY OF THE ECOSPHERE

The famous explorer Captain Scott, on reaching one of the ecosphere's least hospitable areas – the South Pole – exclaimed "Great God, what an awful place." Given the historical shift in the meaning of the word awful from 'awe-inspiring' to 'appalling', we can perhaps be forgiven for questioning whether Scott was referring to the awe-inspiring landscape, the appalling climatic conditions, or both. In describing the whole of the ecosphere as being 'awful' in its complexity, I certainly intend the word to be interpreted in both of its senses. That is, the complexity is so great that it inspires a sense of awe in its would-be students, but also a sense of the appalling amount of work that will be required in order to unravel its many mysteries.

An interesting comparison, which highlights the difficulty with which the ecologist is faced, is that between the number of interacting entities of which his object of study – the ecosphere – is composed and the number of interacting entities in the areas of study of the 'core sciences', namely chemistry and physics. Physicists, as students of the various forms of energy, are perhaps most fortunately placed. Even in the old days of heat, light, sound, electricity and so on, the number of forms of energy was in single figures and the number of possible kinds of interaction between them correspondingly small. Now, in the age of synthesis and unification in physics, the number of 'forces' at work in the universe has been reduced to a mere two or three and the striving for synthesis and unification has still not ceased. Chemists, who focus their attention on matter rather than energy, have a longer list of interacting entities to deal with. This list comes in the form of the periodic table, but although this may look complex at first sight, the number of chemical entities it represents is not too far into treble figures.

Consider now the number of interacting components that constitute the ecosphere. It isn't even clear how this overall system should be broken down into components, but with regard to its biotic 'half' (to which this book is largely addressed), the most obvious breakdown is into species, since species populations are the interacting entities which populate the ecosphere and give it its uniqueness. The number of species of living organisms is impossible to estimate with any degree of accuracy, but most guesstimates that have been made put the overall number – inclusive of plants, animals and microbes – at somewhere between two million and twenty million. The number of potential interactions between species is therefore extremely large even if only pairwise interactions are considered, and virtually infinite if more complex interactions – such as food chains – are taken into account, as they clearly must be.

Of course, the number of potential interspecific interactions may be grossly in excess of the number that actually take place, and there are several reasons why we would expect this to be so. Foremost among these is the fact that species' geographical ranges vary widely in size from those that are 'cosmopolitan' (i.e. more or less world-wide) to those that are restricted to a very small area. Man is an obvious example of the former end of this continuum while the finch *Geospiza fortis* is found only in a single group of islands (the Galapagos) and is therefore an example of the latter. Clearly, two randomly chosen species of very restricted distribution are unlikely to overlap geographically, and cannot therefore interact ecologically, whether as competitors, as predator and prey, or as anything else.

The inability of some pairs and groups of species to interact with each other means that our initial considerations of the number of potential interactions in the ecosphere may represent an overestimate. Nevertheless, there is another phenomenon that needs to be considered which points in the opposite direction (notwithstanding the fact that infinity cannot be an underestimate!) This is the phenomenon of genetic variation within species and particularly the genetic differences between populations of the same species inhabiting different environments.

Unlike a chemical element, which is almost a fixed entity and is restricted at most to a few isotopic states, a species population is a much more variable thing. We have to look no further than our own species to observe clear genetic and phenotypic differences between populations living under different climatic (and other) conditions. And if we do look further afield, we find exactly the same phenomenon – differentiation into more or less well-marked races or varieties – in any species of even moderately wide geographical range. Thus, for example, in comparing two closely related species, we may find that one of them is competitively superior to the other when a particular pair of local populations are compared, yet the superiority may be reversed in other localities because of genetic variation as well as other factors.

Biological Species

Clearly, the foregoing account is based on the assumption that we know what is meant by 'species', but it's worth digressing to consider whether this is actually so. Most people – even those with little biological training – have a rough idea of what 'species' implies, and could readily give a few examples such as *Homo sapiens, Drosophila melanogaster* and the appropriately named *Gorilla gorilla*. These, of course, are all examples from the animal kingdom, and it is here that there is least difficulty with the species concept. Animal ecologists, together with other zoologists, regard as a species any group of individual animals such that interbreeding is possible within the group but not between it and other, similarly defined, groups. This view of the species is often referred to as the 'biological species concept' to distinguish it from the earlier view of the species as a group of organisms delimited from other groups by the greater morphological or structural similarity of its members to each other than to organisms from those other groups. This is the 'morphological species concept'.

There are two obvious advantages of the biological species concept, one of them general and the other more specifically ecological. It's more satisfactory to use a binary variable such as 'do/do not interbreed' to define a grouping of organisms than a continuous variable such as 'degree of morphological similarity'. On the ecological front, where interactions among organisms are a major focus of attention, it's useful to have a species defined as a group within with an additional kind of interaction – reproduction – is possible over and above the sort of interactions (e.g. competition) that can take place regardless of whether the individuals concerned are conspecific.

Despite these clear advantages of defining species as reproductive groupings, vestiges of the morphological species concept persist. The reason for this is that outside of the domain of the zoologist – namely, living animal species – the interbreeding criterion is fraught with a variable array of problems. Many kinds of plant have an uncanny ability to interbreed not just between 'species' (as in the case of the two British oak 'species'), but even between the larger-scale groupings of related species that we call genera (examples of the latter process being found in many cacti). In microbes, the predominant form of reproduction is asexual, and a 'species' therefore consists of a group of isolated clones between which there is very little reproductive cohesion.

Palaeontologists – even those studying fossil animals – encounter another sort of difficulty. Applying the interbreeding criterion at any moment in time is one thing, but applying it in the four-dimensional context of the fossil record is quite another. For example, suppose that the view of human evolution that has *Homo sapiens* as a direct descendant of *Homo erectus* is correct. Could 'wise' humans from today transported back in time by a million years or so interbreed with their 'erect' ancestors? The answer is clearly indeterminate. All we can say is that if transported back a few thousand years we could almost certainly interbreed with our ancestors of that era, while if transported back many millions of years, interbreeding would clearly be impossible. Yet at no point in the continuum of evolving life from (say) the first small primate-like creature to modern man was there any break in the inter-generation link that reproduction provides. So where are the 'ends' of a species in the time dimension? Again, we cannot satisfactorily answer the question.

It might be argued, of course, that the ecologist can evade such questions, and that ecological attention can be restricted to the present – but this is too narrow a view. A whole branch of ecology – palaeoecology – is concerned with attempts to reconstruct past environments, rather than merely past organisms in splendid isolation. The ecosphere, after all, is itself a four-dimensional entity and we should seek to understand its past as well as its present – and indeed to predict its future, with which problem many 'green' ecologists are concerned. Breadth of vision is one of the strengths of ecology in contrast to some other scientific disciplines, and we should resist any attempt to narrow our scope.

Having aired the problem, what's the solution? That is, how do we arrive at a common view of the nature of species in the face of the diversity of difficulties faced by botanists, microbiologists and palaeontologists? Should we simply retreat to a purely morphological approach? The answer to the last question is an emphatic 'no'. What we should do instead is to retain the biological species concept in all cases but to treat it as a probabilistic concept rather than an absolute one. That is, we regard a species as a group of organisms within which the reproductive links are much 'denser' than those between groups. In some cases (plants) the density of links within a group is very great, the density between groups much lower but certainly not insignificant. In predominantly asexual microbes, the density of links within a group is low (but not zero because mechanisms for sexual reproduction are almost always available, even if only rarely utilized), but the density between groups is lower still. In the temporal dimension, there must be a decrease in the density of reproductive links at the 'start' and 'end' of a species' lifespan. In the animal kingdom, we have the special case where many taxa approach (but rarely reach) the ideal of a clear binary contrast of interbreeding or lack of it with which we started. So we have both a general, and ecologically acceptable (i.e. interaction-based) definition. The only problem is that it is probabilistic. Yet in a sense this is not a problem, because almost every ecological concept is probabilistic rather than absolute, and the species concept is therefore useful in providing practice for later probabilistic thinking.

From a Rolled-up Leaf to the Himalayas

Let's get back to the theme of complexity. So far, we've seen that the ecosphere is astoundingly complex in terms of the number of species that can interact with each other – and we now have a reasonable idea of what 'species' means. But there are other aspects to the ecosphere's complexity, additional to those that are species-based, and we need to examine these as well.

Perhaps the most striking of the others is the sheer physical complexity of the abiotic environment provided by the ecosphere, in contrast to the relative simplicity of the physical environment that prevails in interstellar space or in a chemical solution. The prolific interactions between species which we've already briefly considered are embedded in environments that vary in almost every conceivable way. The prevailing temperature may be below freezing or above boiling. (Temperatures of about 350°C have been recorded in ocean-floor 'sulphide chimneys', around which swarms of specially adapted shrimps have been found.) Organisms may be found within solids (such as the soil) or liquids (a pond), or they may be almost permanently airborne. Those that inhabit interfaces between the three physical states (*most* organisms in fact) may be found at all three types of interface.

Within a major habitat type, as defined by its physical nature, there is still quite extreme variation. Aquatic habitats range from the Pacific Ocean to tropical rainpools which, despite a size of only a couple of square metres, may contain tens of thousands of insect larvae. Terrestrial habitats include the world's various 'flatlands' – variously named plains, steppe or savannah – in the middle of which an observer may not see a single undulation in the ground despite looking all the way to the horizon in every direction. Yet they also include craggy mountain ranges such as the Rockies or Himalayas where there is hardly a horizontal piece of ground to be seen.

For many small creatures the 'habitat' is not so much the major abiotic features of the landscape such as a mountain range, but rather a much smaller-scale, and usually biotic, physical structure such as a rolled-up leaf, a piece of rotting fruit or (for some parasitic organisms) the intestine of a mammal. At this level, the number of physical forms the environment can take is almost limitless. Indeed, here we connect back with the 'species complexity' angle, because at this level the physical environment of a small insect (for example) is often the tissue of a larger organism such as a plant or a vertebrate, and is therefore as variable as the array of species within such habitat-providing taxa.

Another complexity of the ecosphere is that the ecosystems of which it is composed often run into each other, with the result that it's difficult to focus in on a clearly defined spatial area which can be regarded as a meaningful study unit. Such a unit should be characterized by the density of interactions between species within it being much greater than the density of interactions between it and other such units. (Note again the probabilistic nature of the argument.)

There are exceptions to the general occurrence of this continuum problem, of course, and the most obvious example is provided by small oceanic islands. Here, the ecologist is almost as well-placed as his physiological or anatomical colleagues in that the boundaries of a small island ecosystem are almost as clearly drawn as those of an individual organism. Consequently, ecologists have invested a disproportionate amount of time and energy studying islands, and indeed an elaborate body of theory known as 'island biogeography' has been developed. This theory – attributable to the American ecologists Robert MacArthur and E. O. Wilson[2] – concentrates on variables such as immigration rates and extinction rates, on how they are moulded by physical variables (such as distance from the nearest landmass), and on how they in turn determine other variables (such as the species richness of the island).

This way of looking at things is equally applicable to what have come to be called 'habitat islands', such as an array of small ponds scattered across an extensive area of moorland (the photo-negative of conventional islands) or a number of small, dense clumps of woodland in a 'sea' of agricultural fields. Again, the great advantage of such systems is that their boundaries are reasonably clearly defined.

Contrast such systems with the coniferous 'boreal forest' that stretches across most of Canada. What, within such a vast tract of the ecosphere, constitutes an individual ecosystem? Clearly, there are no well-defined borders or boundaries, and consequently, it would seem, no separation into ecosystem-type units. Yet defoliating forest insects may have a patchy distribution, and a patch constituting a local population of the species concerned in a particular place will not interact with a comparable patch of an insectivorous bird species a hundred kilometres away. Their potential but not actual interaction would place them in different ecosystems – yet this contrasts with our earlier conclusion that such do not exist.

The problem becomes even worse when lack of physical separation is compounded with variation in the *type* of the system concerned. Picture, for example, the transition between deciduous and coniferous woodland that would be encountered when going 'up' an appropriate latitudinal or altitudinal gradient. Or alternatively, picture the transition between freshwater and marine environments that takes place through estuarine regions. To what extent do species populations extend through such transition-zones? For some species, the answer is known (e.g. the trees themselves in the forest transition), while for others (e.g. the forest invertebrates), it often isn't. A convenient fiction is that the ecosphere is divided into moderately clear-cut, ecosystem-type units separated by narrow transition zones termed ecotones. While no doubt this model is in some cases a reasonable description of reality, it's more often an oversimplified view of the messy continua that are so frequently found.

The upshot of all of this, of course, is that any present-day ecologist who wishes to make maximum progress (and to retain his sanity besides) directs his attention to reasonably small, isolated systems, such as habitat islands. Only when we have reached a deeper understanding of the ecodynamics of such *relatively* simple cases will we be in a position to begin to investigate the additional problems posed by larger-scale systems where different communities merge imperceptibly into each other.

Realms of the World

It would be possible to go on and on describing all sorts of other complexities of the ecosphere, but there seems nothing to be gained from doing so. If I haven't conveyed some idea of the magnitude of the problem facing the ecologist by now, it's unlikely that any list of further complexities will convince you. Yet there is just one other aspect of the ecosphere's complexity that I feel merits brief consideration – partly because it's relatively cryptic from the viewpoint of the untrained observer, in contrast to observations of the physical heterogeneity and sheer species richness of the world, which no lack of ecological training can hide.

This one additional complexity is the existence of six rather clearly delineated (as ecological things go) regions of the ecosphere whose inhabitant arrays of species differ to a large degree from the comparable arrays of inhabitants of the other regions. These important biogeographical provinces are referred to as Wallace's Realms, after the nineteenth-century British naturalist Alfred Russel Wallace, co-founder with Charles Darwin of the theory of natural selection.

The six Realms (see Figure 1) represent divisions of the terrestrial environment, and their boundaries correspond to seas, narrow peninsulas, deserts and major mountain ranges – that is, features that prevent or restrict the movement of many animals and plants. The result of such restriction is that species originating in one Realm are often confined to it. Consequently, after a long period of evolutionary time (such as has now elapsed) the different Realms are characterized by considerably different floras and faunas. The Realms, as currently recognized, are: Nearctic (North America), Palaearctic (Eurasia), Neotropical (South America), Ethiopian (Africa south of the Sahara), Oriental (Indo-China, south-east of the Himalayas) and Australasian. The faunal and floral differences between these Realms are still very pronounced, but are being gradually eroded by man's tendency to transport species of various kinds around the globe, both accidentally and on purpose. A fascinating account of this process can be found in a

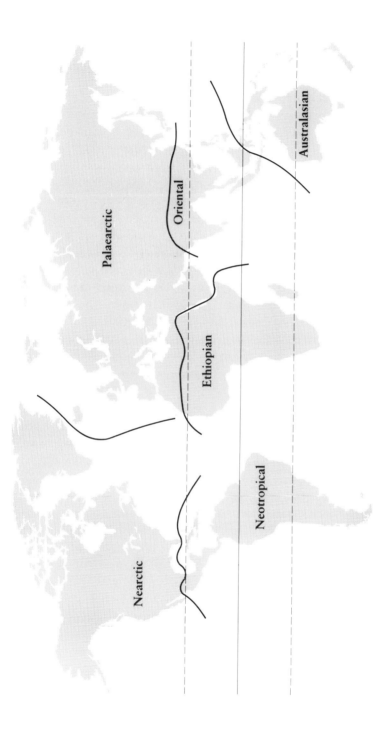

FIGURE 1 Wallace's Realms — the major biogeographic regions of the terrestrial sector of the ecosphere.

book by the 'grandfather of British ecology', Charles Elton.[3]

The existence of this particular ecospheric complexity is actually a blessing in disguise. It isn't wildly easy to explain why this is so, especially at such an early stage in the book, but I'll try anyway. As I mentioned in the first chapter, ecology is primarily concerned with arriving at valid generalizations – general principles if you prefer – relating to the natural world. These general principles are arrived at through building up a series of case-studies of particular species or groups of interacting species living in particular places, and looking for consistencies in the way the systems work, rather than for species-specific or locality-specific details, most of which are eventually discarded. We might, for example, wish to know the typical length of food chains (in terms of number of links) in tropical rainforest ecosystems. A study conducted in the Amazonian rainforest might give the answer to be four. (This number may not be accurate, but that has no bearing on the argument.) After a single study, of course, all we have is an observation whose status is unknown. That is, it might be a very locality-specific statistic, or a very generally applicable one (and therefore one that provides an insight into general ecological principles), but we are so far none the wiser as to where on the continuum bounded by these extremes our observation actually falls.

Further case-studies conducted in other 'plots' of Amazonian forest might well come up with the same answer – i.e. a mean food-chain length of four links – which would give us some confidence that our observation had at least a degree of generality. However, since many of the species in the two study sites will be the same, this is not a very powerful test. In contrast, studies conducted in the Congo Basin or in Borneo (and so in rainforests whose species composition is radically different) provide a much stronger test of the independence of our observation of food-chain length from what can be described as local effects. If in all cases food chains were four links long this would clearly lead us to postulate a general principle at the observational level and also to speculate (and perhaps then experiment) on the stability of food chains of different lengths, to try to find out why in nature the actual number of links kept turning out to be the same. (In reality, of course, the number is variable, though within narrow bounds, as will be seen in later chapters.)

A Word on Tactics

What effect does the ecosphere's complexity have on the way in which we study it? We've already seen two such effects, and they point, in a sense, in opposite directions. In the case of poorly-defined ecosystems merging imperceptibly into one another we try to avoid the complexity and concentrate our attack – for the moment anyhow – on simpler systems such as habitat islands. In the case of Wallace's Realms, in contrast, we can *make use of* the complexity in testing the generality of ecological observations. Perhaps the most powerful and all-pervasive effect, however, stems from the primary aspect of ecospheric complexity with which we started – namely the immense richness of species and the consequent multiplicity of possible (and actual) interactions. This multiplicity prevails even in the simplest of habitat islands and, of course, in all of the biogeographical Realms.

What we're saying, then, is that the interacting plethora of species co-habiting a particular area – a plethora that goes by the name 'community' – presents considerable complexity wherever and however a study area is chosen. It is this complexity that gives rise to what the American evolutionist R. C. Lewontin[4] has aptly described as the 'agony' of community ecologists. That is, the system that is the object of study – the community – is so complex that not only are we largely ignorant of its mechanics but indeed we are not even clear about what are the most sensible questions to ask.

At present, community ecologists are still enmeshed in the iterative process of asking questions, collecting data which may help to answer them, but finding that the 'answer' is that the wrong question had been asked in the first place. However, we have come far enough through this iterative process to be able to state what appear to be the most central problems with a greater degree of confidence than would have been possible a decade or so ago. These problems and their relative priority constitute an ecological agenda, and form the focus of the following chapter.

3

An Ecological Agenda

With the greatest of ease, you could spend a lifetime doing ecological research and get precisely nowhere. Indeed, many ecologists would argue that some of their colleagues have done precisely that. The problem is that faced with the bewildering complexity of the ecosphere which we have just sampled, it's tempting to grab the most convenient community as your study-system, put your head down, measure everything in sight and build up as much data as possible. However, this is a thoroughly misguided approach. Robert MacArthur, the American ecologist whose untimely death robbed ecology of one of its most creative practitioners, reminds us that "to do science is to search for repeated patterns, not simply to accumulate facts." He adds that the best science is "guided by a judgement, almost an instinct, for what is worth studying".[5]

What, then, *is* worth studying in ecology? To come up with a sensible answer to this question we need to look at it from a variety of angles. Ultimately, a practising ecologist needs a very specific answer, in that he needs to single out a particular problem on which he will focus his attention. However, we can only get to that degree of specificity in stages, and it's wise to take a broader view of the question first.

Probably the best starting point in our process of narrowing down from the whole of the ecosphere to what will hopefully be a productive study-system is to address the question of what degree of generality of explanation we are seeking. I've already made the point that ecologists are interested in constructing general theories rather than getting enmeshed in species-specific details, and that that is what distinguishes them from natural historians. MacArthur's comment about searching for 'repeated patterns' clearly reinforces this view, since repeated patterns are what general theories seek to explain. Does that mean that we should strive for maximum generality and minimum detail?

There is a very good reason why the answer to this question should be 'no'. If we try to paint too general a picture, there is a danger that it will be nothing more than a statement of the obvious. The holistic view of ecosystems as a series of 'trophic levels' (plants, herbivores, carnivores, detritovores) which together constitute a system of energy flow and nutrient cycling (see Figure 2) comes perilously close to being just such a statement. Indeed, we could argue that a perceptive caveman would have had a fairly solid grasp of the ideas of trophic levels and energy flow, and that

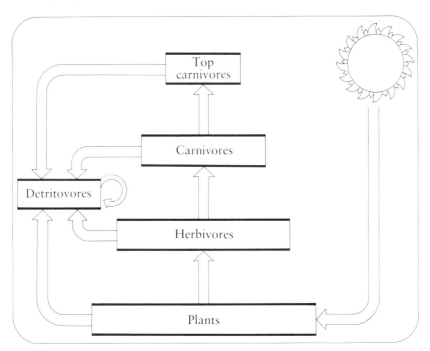

FIGURE 2   The energy-flow pattern in a typical ecosystem. Nutrients follow a similar pattern to energy except that there is an additional arrow from detritovores to plants.

all we have done is to add the jargon – a dubious advance indeed.

What I am suggesting, then, is that ecologists should follow a middle course. We should not allow ourselves either to become submerged in a sea of species-specific facts (the lifespan of badgers, the diet of crayfish, the temperature-tolerance of the blue whale, and so on), or to be content with such a simplistic general picture as the 'energy flow between trophic levels' one. Nevertheless, the next chapter will be devoted to the simple holistic view because, while it is hardly a sufficient end-product of ecological research, it isn't a bad starting point.

From Levels to Chains to Webs

The main problem with the simplistic trophic-level picture, as shown in Figure 2, is that each 'box' is treated as a unit, and only interactions *between* such units are made explicit. This kind of approach manages to hide all sorts of important ecological interactions. For example, one of the most important kinds of ecological interaction – and one which may have a significant influence on community structure – is competition. This may take place between individuals of the same species, or between one species and another, and examples of it may be found in all the major trophic levels, though it's thought to be commoner in some (e.g. plants) than others (e.g. herbivores). Competition involves no direct energy flow from one competitor to another, in contrast to predator-prey interactions. It may take a variety of forms – plants competing for light, birds for nesting sites or carnivorous mammals for a limited supply of prey. All such competitive interactions are hidden from view when the trophic level is the smallest unit into which an ecosystem is divided.

To arrive at a more comprehensive view, we need to make a fundamental switch in the unit of study – from trophic level to species population. We need to picture a community not as a simple array of a few interlinked 'boxes' but rather as an *interaction web* of all the populations which comprise it.

Suddenly, there are a lot of important distinctions to be made. Interaction webs are not the same as food webs, nor are the latter the same as food chains. Both interaction webs and food webs can be broken down into sub-webs, which are something else again.

In order to be clear about the distinctions, let's start at what is probably the simplest – and least realistic – of these concepts, namely the food chain. This has virtually become a household word, and requires little explanation. If a plant is eaten by an insect and that insect eaten by a bird, we have a food chain – a shortish one, in this case, consisting of a mere two links.

According to one way of looking at nature, food chains exist, while a different viewpoint immediately suggests that there is no such thing. Let's deal with these in turn. The 'food-chain-affirmative' view arises from considering what happens to a *particular piece* of plant tissue that gets eaten by a particular insect (for example). If that insect is later consumed by a bird, we have a food chain which can be represented by a sequence A→B→C, where each of A, B and C represent a nameable species. Such chains run at rightangles to the trophic-level 'blocks' found in simple diagrams such as Figure 2 and hence are sometimes described as 'vertical' interactions.

A slightly more sophisticated view immediately reveals the artificial nature of the food-chain concept. Here, we need to return to probabilistic thinking. Suppose we start with the same piece of plant tissue, but *before* it gets eaten. It's likely, in that case, that several different herbivores which are present locally would be able to consume it. Taxonomically, these may be very diverse, and might include several insect species (including the one we started with), several species of slugs, a couple of bird species and a rabbit. Thus from a point of view of what *could* eat tissue from the plant concerned, rather than what *does* eat a particular piece of such tissue, we need several arrows feeding out of 'A' (which represents the plant species) rather than just one. Conversely, since most herbivorous species will be capable of consuming several species of plant, we need several arrows feeding into 'B' (the herbivore). The same applies to most consumer species, regardless of their trophic level. Thus if we want to picture what happens in a natural community in the course of a year (say), we need a food *web*, wherein there is a complex link-up between the constituent species of the system, rather than just a series of separate chains.

While a food web is clearly an advance on the concept of a food chain, it still has a serious failing. That is, like the trophic-level concept, it omits any information on non-trophic interactions such as competition. However, such interactions may easily be accommodated within a web-like structure; and indeed, food webs almost suggest where competitive-type interactions should be expected, because they indicate overlaps between the diets of different consumer species. When competitive interactions are included along with the trophic (feeding) kind, we have what can be referred to as an interaction web, as illustrated in Figure 3. Strangely, while this kind of picture contains more information than any of the others, and can therefore be thought of as the most

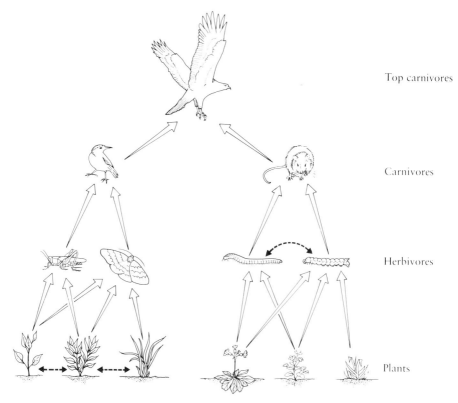

Top carnivores

Carnivores

Herbivores

Plants

FIGURE 3   An interaction web, showing trophic (↑) and competitive (←--→) interactions. Note that dietary overlap *can* but *need not* lead to competition (more on this later), as illustrated by the herbivores here.

satisfactory of our abstract pictures of natural communities, it is something that I have had to invent for the purpose of this book. 'Food chain' and 'food web' I have encountered numerous times in the ecological literature, but 'interaction web' I have never seen.

A community isn't just characterized by one web of the sort depicted, but rather by several, and these may overlap to varying degrees. I have, in common with most other ecologists who portray webs, given an example (in Figure 3) which relates to the macroscopic, live-feeding part of the community – the part that, in a terrestrial system such as a grassland or forest, is most visible. Parasites (often microscopic) and detritovores (often hidden in the soil) also contribute to the overall interaction web of the community; and even without these additional complexities, a number of partially overlapping webs culminating in different 'top predators' may often be recognized. Such components of the overall web are referred to as sub-webs, and these have sometimes been productive study-units, as subsequent chapters will reveal.

Even sub-webs are often too complex to study in their entirety, and a breakdown of the community into smaller components is desirable. An often-adopted strategy – and one that has been highly productive – is to concentrate on the individual interactions of which an interaction web is composed, namely those between a predator and its prey or between one competitor and another. (As we'll see later, other interaction types also exist, such as symbiosis or mutualism, where species interact to their mutual benefit). However, we have to be careful here. If we end up studying such small units, and therefore as few as two species at a time, we must be on our guard against a retreat into the pit of species-specific detail. If birds of prey, insect parasitoids, wild cats and sharks do not exhibit any general 'patterns of predation' but rather have to be described in entirely individual ways, then there is no ecology to be done, only natural history. But as we'll see, such a pessimistic view is unjustified.

The approach that will be adopted in this book, then, is as follows. First, we look at the 'trophic-level overview' to find our bearings. Then we descend to the level of the species population, and look at its growth, its regulation, and its interactions with other such populations. Then we begin to build our understanding of such simple units back up into a more comprehensive picture of whole communities than was possible before we began our descent. The building-up process works in stages. Trophic interactions can be 'stacked' to form food chains, – the vertical dimension of the community. Competition between a pair of species can be extended to take in the many species that comprise a 'guild' (e.g. seed-eating birds) – the horizontal dimension. The two dimensions can then be merged to give the interaction web that essentially *is* the community.

Having arrived back at the community level, but in a more enlightened frame of mind, we can then progress to looking at whole-community properties and processes, such as species-richness and succession. And, somewhere along the line, we can inject a modicum of evolutionary awareness, which adds a longer-term view. Finally, armed with all this information, we can examine man's intentional and unintentional tinkering with ecological systems, and consider the potential benefits and hazards of such tinkering. We shall then no longer be 'ungreens', but we'll be cerebral rather than knee-jerk greens, and therefore, in the long run, more of a force to be reckoned with.

Stability in All its Guises

So far, our 'agenda' has concentrated on the degree of generality we should seek to achieve in our explanation of ecological systems, and on the size of unit upon which study should be based – population, guild, web or trophic level. What we haven't yet addressed is the question of what *property* of ecological systems should provide the focus of our attention. Here, for once, there is near-unanimity in the ecological world – a rare thing indeed. Greens and ungreens, reductionists and holists, more or less everyone, in fact, would agree on the number one choice – stability.

What, then, *is* stability, and why are we all so convinced that it is at the core of ecological research? Let's take these two questions in turn. Stability, in the broad sense, encompasses several quite distinct properties, of which I'll concentrate on the three that are arguably the most important.

The first of these is persistence. A population (or other ecological unit) is described as persistent if it lasts for a long period of time. If, on the other hand, it goes extinct within a few generations of its inception, it is clearly not persistent. Of course, persistence is entirely a relative concept. While extreme cases may be described as 'very persistent' or 'not persistent', most populations are somewhere in between, and so the concept is most useful in making comparisons of one system with another. For example, a tree population in a mature, undisturbed habitat is much more persistent than the population of a 'weed' that rapidly colonizes bare ground and gives rise to vast numbers of high-dispersal-ability seeds before being ousted by superior competitors.

Even the comparative use of persistence has its problems. The tree/weed comparison is unfair for two reasons: first, because the generation times are unequal, and we should really measure persistence in generations rather than units of absolute time such as years; and second, because a more subtle view of our weed population renders it much more persistent than it first appeared. What I'm getting at here is that we can take a broader-scale view of such populations. Rather than consider the dandelions (for example) colonizing a single patch of bare soil as a population, we consider all those within a larger area (say a hectare) as the population. Within that area, various disturbances may give rise sporadically to bare patches, each of which is transient but which conglomerately form an effectively permanent habitat-type which may be inhabited by a persistent, albeit mobile, population. Thus our view of the persistence of a population depends on precisely what we consider a population to be, and the same general message applies to attempts to describe the persistence of other kinds of ecological unit.

We turn now to the second of the three main stability properties, which is – wait for it – stability. We have an unfortunate problem here. 'Stability' is used in both a broad and a narrow sense, and the amount of confusion that this has generated in ecology is immeasurable. I propose to solve the problem by only using 'stability' – from here on – in its strict sense, and using another label (which I'll introduce shortly) for general-sense stability.

Strictly, then, an ecological system can be described as stable if it has an equilibrium value, level or state to which it will return after being perturbed away from it. For example, a population of an insect species is stable if sudden increases in population size (e.g. due to immigration from neighbouring populations) are followed by a 'decay' back to the pre-increase level, and sudden depletions (e.g. due to a prolonged period of drought) are soon compensated for by population growth back up to the same level – the equilibrium level, as it's called. As with persistence, the term stability can be applied to a variety of ecological units. For example, we can think of the species richness of a whole community as being stable or otherwise – depending on its behaviour following perturbation – in just the same way as we think about the stability in numbers of individual species populations.

One of the advantages of stability over persistence is that it's an absolute, not just a relative, concept. The size of a population either is or is not stable, and we can therefore say something meaningful about individual systems, as well as making contrasts between them. It must be admitted that some ecological systems exhibit various kinds of cyclic behaviour, which are difficult to characterize from a stability viewpoint. We'll look at some of these later.

While stability and persistence are distinct concepts and it is necessary to be clear about this, there is nevertheless some interconnection between the two. For example, if a population is stable, we'd expect it to be fairly persistent, while if it is unstable we'd expect it to be much less so. However, this apparently simple relationship between stability and persistence is complicated by the third of the three 'properties' to which we now need to turn.

This is the property of robustness, and it is without doubt the most difficult of the three to explain or to comprehend. Again, an example may help, so let's return to our stable insect population. Suppose, now, that the source of the stability is a limited food supply – one that will support the equilibrium population size but no more. (Limited food supplies provide the most obvious, but by no means the only, source of population stabilization, as we'll see later). Thus the supplemented population declines and the depleted one grows because there is respectively too little or 'too much' food.

Robustness concerns the persistence of the stability! For example, our insect population's property of stability may remain in the face of various changes in the nature of the community in which it is embedded – invasion of the community by a novel predator or competitor, for example – or alternatively such changes may destroy the stability, in which case the population may well go extinct. Since most ecological systems are prone to invasions and other disturbances, populations will be very persistent only if (a) they are stable and (b) their stability is reasonably robust.

So much for our three properties individually. How do we refer to them collectively, given that the 'stability' label has been usurped by one of them? The most appealing answer to this question – to my mind anyway – is that when we wish to talk about the 'stability' of ecological systems in a vague and general way, we refer to the 'balance of nature'. Thus persistence, stability, and robustness are three of the central strands in the balance of nature, which seems a reasonable proposal.

There is an unexpected and beneficial side-effect of this choice – we can redefine ecology! The standard definition is that ecology is the study of organisms in relation to their environment, which is pretty comprehensive but rather uninspiring; and while it clearly isn't wrong, it doesn't give much of the flavour of ecology as distinct, for example, from natural history, for which the same definition would suffice. Let's therefore discard this definition and replace it with a better one. In 1964 the British ecologist C. B. Williams wrote a marvellously-titled book, *Patterns in the Balance of Nature*.[6] And what better definition for ecology than 'the study of patterns in the balance of nature'? This at once conveys both the centrality of balance/stability in all ecological endeavour and the emphasis on repeated patterns rather than isolated 'facts' that we saw earlier in the comments of Robert MacArthur.

33

Balance at the Top of the Agenda

Since we now have a reasonable feel for what is meant by the 'balance of nature', in terms of its constituent themes of persistence, stability and robustness, it's time to turn to the question of why this forms the focus for most ecological study. After all, the balance of nature is not synonymous with nature itself. We could, for example, take a very 'structural' approach, akin to that of the anatomist in relation to individual organisms. That is, we could take a community and 'dissect' it, so that we knew in detail the precise locations of its constituent populations. Insect species X might be found under the bark of tree species Y, whose dead leaves may harbour colonies of fungal species Z. If we took a reasonably well-defined community unit, such as a small piece of woodland surrounded by fields – a habitat island, as we've seen – a very detailed picture could be built up.

Already, however, we begin to get an inkling of the problem. Communities are not as tightly controlled or 'homeostatic' as the bodies of individual animals or plants. An anatomist can amass an exceedingly large body of data on the structure of a brown trout (say), and this will be applicable both at many points in the lifespan of an individual and to most individuals of the same species. The ecologist is hardly so fortunate. Similar-sized clumps of woodland are not replicates of each other to anything like the same extent as are conspecific individuals. Even within a single clump, by the end of the study period (of perhaps a couple of years), much of the information would no longer be accurate – for the very reason that there are limitations to the persistence and stability of species populations. Since there is no point in compiling information that is unique to one clump of woodland, let alone information that does not even apply to that clump a short time on, we are essentially forced to consider 'balance' questions at the outset in addition to anything else (like structure) that we wish to study.

Quite apart from the fact that any ecologist sufficiently blinkered to attempt a 'balance-free study' would soon be forced out of his tramlines, there is a more positive reason why questions about the balance of nature have been so central in ecological endeavour. In other words, there is a good reason why the vast majority of ecologists are not blinkered in this way in the first place. Here, we come back full-force to the emotion of wonder with which this book started, and which, as we saw, lies at the core of ecology.

Many people – Charles Darwin among them – have wondered at the extraordinary capacity for lack of balance in the ecosphere. Any species could cover the entire surface area of the planet several miles deep given far fewer generations than have been available to them – yet they do not. Populations of astute, rapidly moving and rapidly breeding predators could eat their prey to extinction and then go extinct themselves – but they rarely do so. Competition between closely related species for a limited food supply usually leads to extinction of one of them when simple population experiments are conducted in the laboratory – yet such extinctions are much rarer in nature. Plant populations could be decimated and land laid bare by over-enthusiastic herbivores, yet in natural systems phenomena such as locust plagues are the exception to the 'rule of the mild-mannered herbivore'.

It's a source of wonder to most ecologists that natural communities appear so relatively balanced, given the potential for imbalance of many very dramatic kinds. How such balance is brought about in the face of forces tending to destroy it thus thrusts itself forward as a – almost certainly *the* – key question. So we do not have to agonize over what is the central issue in ecology, because it virtually throws itself at us – hence the near unanimity on its importance.

Our 'agenda' is now almost complete. We know what properties of ecological systems ecology should concentrate upon, the degree of generality of explanation to which it should aspire, the sort of small ecological units upon which initial study should be based if it is to be productive, and the larger units – communities and ecosystems – whose structure and function it is ultimately seeking to explain. That is, we have finally reached the point of the practising ecologist: we could choose a study-system, begin to work on it, and be confident that we were participating in the search for repeated patterns, and not just accumulating data for the sake of it. We'll examine at the results of such studies already undertaken by practising ecologists in many of the ensuing chapters. But first, we'll look briefly at the simple 'holistic' view of ecosystems. This serves to paint a broad picture of the ecological scene, before we begin to fill in the details of the 'players'.

4

A SIMPLE HOLISTIC VIEW

Having been earthbound for the past couple of chapters, it's time to return to space. Suppose that, on your 1000-planet sampling mission, you have now reached 'planet 572'. By this stage, you're fairly blasé about the landing, and you've a fair idea of what to expect – another barren landscape of rocks, dust and 'staticness'. As usual, nothing whatever seems to be happening. However, this is something of an illusion, because – unless you make a night-time landing – you are being bombarded by a continuous influx of energy from whatever star is your local sun. The intensity of this energy will depend on how far 'in' or 'out' planet 572 is within its particular solar system. If it's an innermost planet (comparable to Mercury in our own system) the radiant energy you receive may well fry you and melt your spacecraft. Assuming you're sufficiently prudent to avoid such environments, the intensity of the incident energy will vary from 'middling' (as on earth) to pretty dilute, in the outermost planets. Yet even on planets most distant from their respective suns, the energy input is far from negligible – it's merely small in comparison to their inner neighbours.

What happens to this vast energy input experienced – to varying degrees – by all planets? In order to answer this question, we first have to become familiar with a few basic facts about energy in general. One essential fact is that energy cannot be created or destroyed. This is acknowledged in the First Law of Thermodynamics, probably better known by its alternative and more informative title – the law of conservation of energy. Of course, since Einstein, we have to consider matter as a form of energy for this law to be valid. Hiroshima and Nagasaki are terrible testaments to the amount of energy that can be released from a very small amount of matter – Einstein's e = mc². Indeed, our terrestrial ecosphere could be annihilated by a computer virus, because of the vast amounts of destructive energy that the computer-controlled nuclear 'defence' systems of the superpowers could unwittingly release, according to the same formula.

Although energy can neither be created nor destroyed, it's constantly being converted from one form to another. We're all familiar with examples of the different forms of energy – heat energy from fires, electric energy from power stations, the kinetic energy of a railway locomotive and so on. That energy interconversions occur is also a familiar fact – from electricity to heat in the case of an electric fire, for instance. However, what is rather less obvious is that there is a pattern of energy conversions wherein 'high-grade' energy is converted – in various ways – to its 'low-grade' equivalent.

Let's think about this proposal for a moment. The primary kind of high-grade energy in the universe is light, which provides a sort of 'energetic starting point'. Light can be converted into all other sorts of energy. For example, consider the following sequence: (a) light to fossil fuel via photosynthesis; (b) fossil fuel to electricity via an oil- or coal-fired power station; (c) electricity to heat via an electric fire. Heat is the end-point in energy conversions, and ultimately any energy 'package' starting from the sun (for example) ends up in that form.

I should stress that it is not impossible to go 'backwards' in particular parts of the sequence just described. For example, light bulbs produce light from electricity. However, all energy interconversions are of limited efficiency and involve the loss of some energy as heat. Of the total energy emanating from a typical light bulb, less than 10 per cent is light, with heat accounting for the other 90 per cent or so. Overall, then, the pattern of energy interconversions in general can be represented as in Figure 4, where the intermediate stages A, B and C (the number is arbitrary) represent different types of energy in different examples.

We can now go back to our task of analysing the fate of the light energy arriving at planet 572; and let's suppose, in order to provide a mental picture of the planet's surface, that planet 572 happens to be very similar to Mars, whose landscape is reasonably familiar to us from the *Viking* photographs. We know that the incoming energy is light – but what interconversions (if any) occur, and in what form does the energy that arrives at the planet leave again?

Actually, the answer to this question is quite simple. First, some of the light that enters the atmosphere is simply redirected and leaves again in the form in which it arrived. After all, we couldn't see Mars as a bright object in the night sky if it didn't reflect incident sunlight – planets certainly have no light source of their own. Second, some of the energy is converted to heat. During daytime the incoming light excites the molecules of the ubiquitous rocks and dust, thereby warming them up, while at night these warmed bodies lose their heat back into space. (Cloud cover

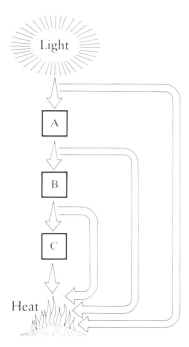

FIGURE 4   A generalized sequence of energy conversions, illustrating the downgrading of high-grade to low-grade energy via a series of intermediate types (A, B, C), whose number and nature will vary from one system to another.

reduces this heat loss, which is why cloudy winter nights on earth are less likely to be frosty than clear ones.)

Although we could analyse the energy relations of planet 572 in more detail, nothing much would be gained from doing so. Essentially, light comes in, light and heat go out. Some of the energy has been 'downgraded' at the planet's surface (and in its atmosphere), and some has not. The total amount leaving (in whatever form) equals the total amount of light energy received, providing we choose an appropriate timespan for our measurements.

The Green Machine as an Energy-Flow System

While the energy relations of a lifeless planet can be seen to be rather simple, the same cannot be said of the earth – or of any other ecosphere-bearing planets, should such exist. Here, the pattern of energy flow is much more complex. We'll concentrate on that part of the energy that is absorbed by autotrophs ('self-feeders', i.e. plants), because that's the most ecologically relevant. It must be stressed that this is a tiny minority of all the solar energy that arrives at the earth's surface; but its importance vastly outweighs its insignificance in proportional terms. Without the solar energy that is acquired by plants in photosynthesis, there would be no ecosphere to talk of – at most a motley collection of unicellular forms based on a limited supply of chemical energy – and certainly no ecologists to observe it.

What happens to the solar energy that is captured and converted to 'biological energy' by plants was depicted in a simple way in Figure 2. Essentially, the ecosystem can be thought of as an energy transfer machine – the Green Machine of our title. With each consumption – of plant by herbivore, herbivore by carnivore, and carnivore by 'top carnivore' – a fraction of the energy in the lower trophic level is conveyed to the level immediately above it. At each inter-level transition much of the energy is wasted, and eventually we reach a 'top' level that contains so little energy that it cannot sustain another one.

So far, all we've done is to acknowledge the existence of trophic levels (which can be up to about five in number) and to make the obvious point that there's a sort of one-way system for energy flow among these levels. This in itself is not very informative, and it amounts to what I've rather facetiously called the 'perceptive caveman's view' of the ecosystem in Chapter 3. Is it possible to progress beyond this simplistic vision, without ditching the energy-flow-between-trophic-levels approach?

Different ecologists will give you different answers to this question. My own view is that the correct answer is 'yes, but not very far'. I'll spend most of this chapter outlining the progress it is possible to make without a shift of approach, and most of the rest of the book discussing the alternative approach – based on species populations – which I think will get us further quicker in our understanding of ecological systems. If that statement tempts you to skip the rest of the present chapter, resist the temptation. The limitations of one approach to a subject often illustrate the rationale behind adopting another one much more clearly than an isolated description of the chosen approach alone.

Pyramids and Ecological Efficiency

The steps forward that can be made from a starting point of trophic levels lead in what at first sight look like two quite separate directions – one relating to the structure of ecosystems and the other to their function. As we go further into these, the terminology will reflect the structural and functional themes. What could be more structural than a pyramid, or more functional than a measure of efficiency? Yet, as we'll see, the two 'directions' are in fact closely related, and the appearance of a split is only illusory.

Let's start with the functioning of ecosystems and the concept of ecological efficiency. In order for this 'efficiency' to be meaningful, it helps if you think of an ecosystem as a machine that was actually designed as an energy transmission system. Thus the plant/herbivore link could be described as less efficient than the herbivore/carnivore link if a smaller proportion of the energy available at the base of the link is transmitted through it.

One of the difficulties with this way of viewing ecosystems is that efficiency is a very subjective, viewpoint-dependent concept. In defining efficiency the way we do, we take (appropriately enough) the viewpoint of a top predator. For such a beast, maximal energy transfer through the system is maximally efficient from its point of view. However, from a plant's perspective, the ideal ecosystem is one with the lowest possible (i.e. least 'efficient') energy transfer capacity. So while the efficiency concept is useful for certain purposes, it's worth bearing in mind that ecosystems are not really engineer-designed machines with a single 'task' but rather are heterogeneous collections of interacting species, each of which would take a rather different view of what is 'efficient'.

There are, in fact, several different efficiency concepts (all top-predator orientated) in current use in ecology, but I see no point in surveying these exhaustively – there's a wealth of standard texts from which to choose for anyone who would like such a survey. Instead, let's concentrate on one particular efficiency measure and see where we can get with that. An obvious choice is the energy input to a particular trophic level per unit time divided by the equivalent input for the level beneath it. This particular efficiency measure (which is undefined for plants) has, like alternative measures, a scale of possible values from 0 to 100 per cent. In practice, of course, the range of *actual* values that we find is much narrower.

As we saw earlier, the task of ecologists is to search for repeated patterns in the natural world rather than to accumulate facts; and to attempt to explain these patterns not in a piecemeal way but by reference to general principles whose validity extends beyond the peculiarities of any set of local conditions. The search for patterns can be applied to any level of ecological unit – from population to ecosystem – and to any property (stability, persistence, etc.) that can be measured. In the present context, then, we must look for repeated patterns in efficiency, and the relevant unit is the trophic level.

It's all very well to keep making the contrast between repeated patterns and isolated facts, but we must acknowledge that until many measurements have been made no meaningful patterns will emerge. Initial efforts by ecologists in the area of ecosystem efficiency were thus put into simply measuring the efficiencies of lots of inter-trophic-level links in lots of different ecosystems. Such *directed* data accumulation is perfectly valid (and indeed essential) as a preliminary stage before any kind of generalization can be attempted.

There is now enough data on efficiency in ecosystems to arrive at some simple generalizations, and it is to these that we should turn our attention. So what are these generalizations? First, it appears that there is a 'typical' efficiency for energy transfer between trophic levels of about 10 per cent. Second, there is some variation around this typical figure, with most estimates varying from about 'half-typical' (5 per cent) to about 'double-typical' (20 per cent). Third, given this range of variation overall, there is a tendency for plant–herbivore links to have lower efficiencies than links between herbivore and carnivore or between one carnivore and another.

We have, therefore, at least two patterns. One is the overall 'clustering' of efficiencies within a fairly narrow band on the spectrum of all possible efficiencies, with the cluster of observed values being characterized by both its average and its variability around the average. The second pattern emerges from analysis of the variability, and takes the form of some *types* of link having generally lower efficiencies than others.

Given the sought-after 'repeated patterns', how do we explain them? It must be admitted that in this particular case we cannot yet give more than a vague and tentative answer. With regard to the limitation of ecological efficiencies to a narrow band fairly far down towards the bottom of the scale, it can be speculated that this represents the limit to the effectiveness of organic machines designed by mutation and natural selection. With regard to the greater ecological efficiency of carnivores over herbivores, it seems likely that this is due to the relative indigestibility of many plant tissues. Energy used in the construction of leaves by a tree is, for example, fairly readily exploitable by herbivores, but that used in construction of trunk and branches is much less so; and even in leaves, the tough cell walls of plants are more difficult to break down than the flimsier membranes around the cells of animal tissues.

So much for efficiency – and now to pyramids. Ecologists have used this term to describe the 'shape' of a stack of trophic levels. By shape, in this context, I mean the pattern that a trophic-level diagram assumes when, instead of representing each level as an equal-sized block, we adjust the block sizes to reflect the relative amounts of matter or energy that is contained in, or flows into, each level. The predominant shape of such trophic-level stacks is as shown in Figure 5 – hence the use of 'pyramid'.

One of the measures which can form the basis for a pyramid-style diagram is the total amount of organic matter – the biomass, as it's called – that is present in each trophic level at a particular moment in time. Suppose, for example, you are working on a deciduous forest where various insect species graze on the trees' foliage, a number of species of insectivorous birds consume the insects, and a single bird-of-prey species constitutes the top predator. If it were possible to determine the biomass either in terms of grams of organic matter or its stored-energy equivalent, you would almost certainly reveal a pyramidal structure in the system. That is, tree tissue far outweighs insect tissue, and so on up the food chain to the top predator, whose biomass would be the smallest of all.

Although a pyramid is the usual shape such studies reveal, it is not, contrary to what might be thought, a logical necessity. Exceptions are known where inverted pyramids and spindle-shapes (among others) are found. An example is provided by the

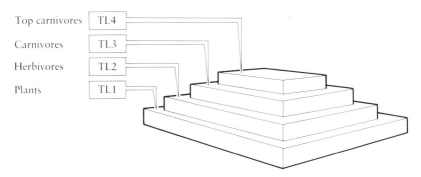

FIGURE 5   An ecological pyramid. Here, each trophic level (TL) is represented by a 'block' whose size is proportional to some measure of matter or energy. (Precisely *which* measure is important: see text.)

Top carnivores | TL4
Carnivores | TL3
Herbivores | TL2
Plants | TL1

plankton, where zooplankton biomass can exceed that of the phytoplankton they consume. How can this be possible, given what we have seen about limitations on the efficiency of energy transfer through inter-trophic-level links?

This is a case where what appears to be the most obvious answer is the wrong one. We might argue that the problem arises because we are measuring the amount of matter, not energy, and there is not a 1:1 correspondence between the two, since some types of matter are more 'energetic' than others. While this is true, it is still possible to get non-pyramidal shapes even if we convert all measurements of biomass into their equivalent energy contents (and so end up with measurements in calories rather than grams).

The solution to our riddle involves what is perhaps a more subtle contrast than that between matter and energy – namely the contrast between an instantaneous or 'snapshot' view of a system and a more leisurely, time-extended view. A measure of the amount of biomass in a trophic level at a given instant in time has something in common with a photograph of a bird in flight: both are accurate representations of the prevailing state of affairs at a particular instant, but both fail to provide any direct indication of temporal changes. In the case of the biomass measurement, there is no indication of the rate of addition of new biomass (production) or the rate of removal of already-existing biomass (by consumption and death). Yet such omissions are odd, since the amount of biomass existing at a given moment of time is due to the balance between these two rates. It's possible to have X grams of biomass with very slow production, consumption and turnover. But equally, we could end up with the same X grams at any instant in time yet turning over rapidly as a result of massive production and consumption rates.

We can now see the solution to the apparent paradox of a higher trophic level sometimes containing more stored energy than the lower one it feeds from, despite the known limitations in energy transfer between such levels. And having seen the solution, it's sensible to incorporate it into our trophic-level overview, by quantifying the size of trophic-level blocks in terms of energy input rather than instantaneous energy content. If we take this approach, then trophic-level diagrams *must* be pyramidal, and our picture of the structure of ecosystems conforms with our picture of their functioning, in terms of limited efficiency of energy transfer. We thus have an overview of ecological systems that is both self-consistent and a reasonable (though highly simplified) representation of what is actually happening in nature.

There's just one proviso. At an earlier point in this chapter I used the rather throwaway phrase "providing we choose an appropriate timespan for our measurements". This is a proviso that could be applied to rather a lot of ecological research activities, so it's worth looking into it briefly in one particular context, to elucidate the nature of the problem. In the present case, where we are looking at energy flux pyramids, it actually *is* possible to have non-pyramidal shapes if we make measurements of relatively short timespan. Let me give a slightly silly example to illustrate the point. During night-time, the rate of flow of energy into the autotroph level in a tropical rainforest is essentially zero. Yet many nocturnal creatures are out and about, feeding from the foliage or, if they are nocturnal carnivores, feeding on the night-time browsers. An energy-inflow 'pyramid' based on such data would clearly depart from a truly pyramidal shape. (It might resemble a pyramid on an infinitesimally thin stilt.)

Problems involving the timespan of measurements don't only beset those who are stupid enough to restrict their measurements to night-time. The problem may occur on a seasonal as well as a daily timescale. Indeed, plankton biomass pyramids, which I noted earlier could be inverted ones, do in some cases flip-flop from one pattern to the other on a seasonal basis, and it's conceivable that, even if converted to energy-flow format, they would still do so. Clearly, the take-home message from a consideration of timespan problems is that measurements must be made on a sufficiently long-term basis that we get a picture of the steady-state dynamics of the system rather than some brief, temporally localized 'blip' in such dynamics. Precisely what length of time this will work out at would be expected to vary between one system and another.

A Middle Way for Ecologists

To summarize, then, the energy-flow-between-trophic-levels approach to ecology presents us with a small number of inter-related general patterns. These include both functional patterns described in terms of efficiency and the structural pattern that takes the form of a pyramid. I said at the outset of our energetic approach to the ecosphere that the structural and functional themes would eventually merge and be seen to be different aspects of the same phenomenon. We have now reached the point where this merging is apparent. Given a limit to the efficiency of energy flow in ecosystems, pyramidal structures inevitably emerge, pro-vided they are based on the right variable. And we have now seen what this variable is – steady state energy flux.

At this point, various additional lines of enquiry suggest them-selves. For example, if plant-herbivore links really are less efficient in general than herbivore-carnivore links, and if such a trend were to be continued up through the various levels, then pyramids should be concave (while if the reverse were true they might be convex). Some types of ecosystem might repeatedly give different shapes of pyramids from others. Finally, given other as-yet undiscovered ecospheres, the universal applicability of the law of conservation of energy should ensure that the pyramidal structure revealed for our terrestrial systems would apply to their distant equivalents too, despite their being composed of very different life-forms.

Fascinating as such questions may be, we will now turn our attention to an alternative approach to ecology. Why this should be necessary is perhaps best illustrated by focusing on what has happened to our central area of interest – ecological balance. This core area of concern is less apparent in the trophic-levels approach than in others, and there is a clear reason why this is so. Basically, trophic levels represent very 'gross' units, each composed of a great many species populations. There could be multiple extinc-tions of such populations, yet the trophic levels would function much as before in the overall ecosystem machine. Indeed, the populations most prone to extinction will be those with the least influence on the system's energy flux, and, given the limitations to our measurement of such flux, they could come and go completely unnoticed.

This line of reasoning brings us back to the 'middle way' that I am proposing for ecology: a compromise between the excessive detail of the naturalist, where every local situation merits its own individual description, and the oversimplicity of trophic levels, where much valuable information is lost by the pooling of populations into giant blocks. The next couple of chapters provide a gearing up for this middle way.

5

Towards Reductionism: The Case of the Carnivorous Cow

When it comes to adopting philosophical stances, most ecologists vote with their feet. That is, most would not consider themselves philosophers (though some do), but by embarking on a particular approach to an understanding of the balance of nature, they are unwittingly supporting one or other of a pair of diametrically opposed philosophical views about how complex systems ought to be studied. One of these is the holistic approach, where the whole system is considered in its entirety from the outset, and any attempt to fragment it to facilitate study is regarded as at best irrelevant, and at worst seriously misleading, because the whole is seen as more than the sum of its parts. If that is true, then a detailed study of the isolated parts may well tell us nothing. The energy-flow-between-trophic-levels approach, outlined in the previous chapter, is an example of the application of a holistic stance.

The alternative strategy is the reductionist one, in which a system is broken down into its components, and each of these is studied in detail before we attempt to synthesize our knowledge and to build back up to an understanding of the overall system with which we started. Of course, in doing this building back up, we may discover that the whole is indeed more than the sum of its parts, but the underlying philosophy of the reductionist approach is that study of the parts will have got us some of the way towards an understanding of the whole, rather than leading us off in some completely different, and unhelpful, direction. A study of species populations and their interactions represents a reductionist approach to the study of complex ecological systems.

As we saw in the previous chapter, ecosystems have a 'holistic' structure, which comes in the form of pyramids, and a 'holistic' functioning, namely energy flow between trophic levels, the limited efficiency of which is the source of the pyramidal structure. The reductionist approach to ecology, which we'll look at shortly, deals with such matters as cycles in population numbers and the dynamics of predator/prey interactions. Since such interactions form the basis for energy flow between trophic levels, it might seem reasonable to expect that patterns of structure and function could be revealed at the lower level that would complement those at the higher one – that is, holism and reductionism would be complementary rather than diametrically opposed to each other.

The relationship is, however, not quite so cosy. One of the problems that immediately arises is that studies of predator/prey interactions reveal that there is no such thing as a trophic level, or, to put it more mildly, that trophic levels are at best a very great simplification of reality. This conclusion arises from a study of the diets of predators. (I use predator here in the broad sense of 'consumer', thus including both so-called herbivores and carnivores.) Predators range in their breadth of diet from those that consume a single species of prey (monophagous) to those that consume a wide variety of taxonomically disparate prey species (polyphagous). The vast majority of predators, however, are polyphagous to some degree. Often, the diets of such creatures include organisms from more than one trophic level.

Examples of this are easy to come by. Some bird species will eat plant tissues (nuts, berries or buds) but will also consume small invertebrates such as insects. The insects consumed may be herbivorous or may themselves eat other small invertebrates. The bird is therefore at once a herbivore, a primary carnivore, and a top carnivore; that is, it occupies three different trophic levels. Where, in that case, do we place it in an ecological pyramid? There is, of course, no sensible answer to this question, which leaves us with the uneasy feeling that perhaps pyramids and trophic levels are not just simplifications of reality but – in some systems at least – are actually distortions of it. Before we come to such a drastic conclusion, we should seek to determine how commonly organisms of unclear trophic-level status are encountered. After all, we can rarely seek universal truths in ecology, where – unlike physics – there are exceptions to every rule.

While I deliberately chose a very hard-to-classify organism to attack the trophic-level concept, consideration of other examples shows that this concept really is very shaky indeed. Cows are a classic example of a herbivore. Yet cows, commonly thought of as eating 'grass', in fact consume many species of grasses, many species of other herbaceous plants that are often found intermixed with grasses, and many species of invertebrates – both herbivores and carnivores – that live within the vegetational matrix. In other words, cows eat entire communities. Cows therefore occupy at least three trophic levels – including (in the case of dairy cows) that of top predator.

There are, of course, some bastions of selectivity in the trophic world. Sparrow-hawks and other birds of prey are not renowned for browsing on grasses while waiting for an appropriate item of avian prey to appear in their neighbourhood. Yet even if the sparrow-hawk was totally monophagous (which it isn't), it could still be of unclear trophic-level status if the species upon which it feeds is like the nut, berry, bud and insect-eating generalist with which we started. That is, no species which consumes any prey of unclear trophic-level status is itself easily classifiable in this respect. And if it consumes several such species of prey (a common situation), then the problem is further exacerbated.

Our final retreat, then, is to plants. Surely at least here we find no difficulties in assigning a trophic level. It's certainly true that the difficulties are fewer, but they're by no means absent. The 'Venus fly-trap' is a well-known exception to the rule of the totally autotrophic plant. What is less well known is that the species of the fly-trap group are vastly outnumbered by other, less conspicuous groups of partial carnivores in the plant kingdom. These include sundews, where the prey are trapped in sticky droplets of fluid, pitcher plants, where the prey fall into a water-filled pool, butterworts, which have sticky leaves like flypaper, and bladderworts, where the prey fall into little pits in the leaves. In all cases the prey are small invertebrates – mostly insects – and once trapped they are digested and the digestion-products absorbed by the plant. Each of these groups of partially carnivorous plants contains many species, and as well as being jointly autotrophic and heterotrophic, they are (in their heterotrophic guise) capable of consuming animals which are themselves from different trophic levels. These plants, then, have an interesting trophic-level 'profile'. They jointly represent level 1 and level 3 upwards, but, since they don't consume other plants (except, perhaps, for a few wandering pollen grains), they have a gap in their activities at level 2 (herbivory). So they are easier to classify according to the trophic

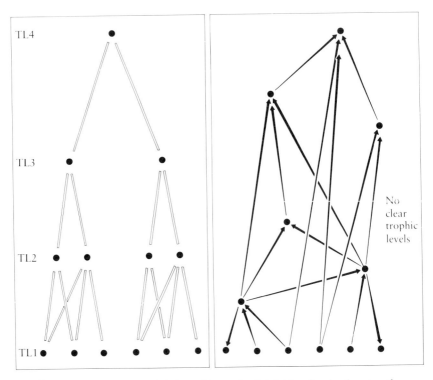

FIGURE 6   A stratified (*left*) and irregular (*right*) food web, showing the dependence of trophic levels (TL) on stratification. Note two 'backward' arrows in the irregular web, indicating carnivorous plants. (Each solid circle represents a species.)

level they *don't* contribute to than according to those they do.

Where, you might well ask, is all this trophic-level-bashing leading us? For once, we have a question with a simple answer – it's leading us back to food webs. It's universally true (despite my earlier lament about the lack of universal truths in ecology) that natural communities have a web-like pattern of feeding interactions. *Sometimes* these may almost simplify into chains, and *sometimes* the webs may be of such a form ('stratified') that reasonably clear-cut trophic levels emerge. But *always* there is some pattern of feeding interactions, and since 'web' covers almost any such pattern, food webs are universal. Figure 6 shows both a regular, stratified web giving trophic levels, and an irregular web (undoubtedly more common) in which no clearly defined levels are found.

Populations – the Molecules of the Ecosystem

We've now seen two advantages of the reductionist, population-based approach to ecology over the holistic, energy-flow-between-trophic-levels approach. First, in the holistic approach, information on individual species is lost, and with it information on that most focal of ecological concepts – stability. (Of course, we can call an energy-flow pattern stable, but this could easily mask instability and loss of the populations of individual species.) Second, we have now seen that trophic levels – the basic units of the holistic approach – are abstractions which often have little correspondence with reality. Why should we deal with a type of system component that is both artificial and uninformative? Clearly, the answer is that we shouldn't, which brings us to where we've been heading for some time: the 'population dynamics' approach to ecology.

In this approach, the community is broken down into its smallest unit – the species population – rather than into just a few massive blocks. In an earlier book, *Theories of Life*,[7] I put forward the idea of the 'two biologies' – the biology of organisms, where the most reductionist approach focuses on molecules, and the biology of ecosystems (i.e. ecology), where the most reductionist approach focuses on populations. Or, to put it another way, populations are the molecules of the ecosystem (and individual organisms its atoms).

As well as the shift in the unit being studied, we also need to make a shift in the main variable that is being monitored. Although it's possible to look at energy flow at the population level, most population ecologists look instead at changes in population numbers. We look at birth and death rates, rates of predation and so on, in terms of numbers of individuals added to, or removed from, the population. This is more informative than recording only energy flow: it's possible to calculate energy flow from detailed population dynamics data on predation, but not vice versa (because, for example, individuals of different lifestages in a population may have different energy contents, thus making it possible to arrive at the same energy flow in different ways). Also, the energy-flow approach is entirely inapplicable to non-predatory interactions at the population level, such as mutualism and competition.

We've now reached lift-off point for our entry to the reductionist, population-based approach to ecology. That means that we can switch from a negative to a positive view. Instead of trophic-level-bashing, we can now turn to look at the great potential of the population approach, and in particular the progress we should be able to make, when adopting this approach, in understanding that most central of ecological concerns – stability.

The Ecological Instability Shop

I now want to introduce you to what I sometimes call 'the ecological instability shop', otherwise known as the laboratory. I use this label because experimental laboratory microcosms of predator and prey or of one competitor and another are notorious for their tendency towards instability and extinctions. In contrast, natural systems have a much greater tendency – at least when left alone by man – for their consituent interactions to be stable. What we have to get at, as ecologists interested in what causes stability in nature, is why the laboratory microcosms differ so markedly from their natural counterparts. If we can solve this problem, then we'll have achieved a major insight into the balance of nature. In doing so, we shall put ourselves in a much better position to assess the potential for de-stabilization of ecological systems inherent in various human activities. We shall thus have a strong basis from which to promote the 'green' view. As usual, ungreen ecology is pre-green (and pro-green) rather than anti-green.

I'd like to illustrate the instability of laboratory-based ecological systems by drawing on the extensive work of the Russian ecologist G. F. Gause, who performed numerous fascinating experiments on protozoans, which are small, unicellular 'animals' (see Figure 7). Although these experiments were conducted in the early 1930s (and are described in Gause's *The Struggle for Existence*, published in 1934[8]), they are as relevant today as they were then; and Gause's work continues to gain attention in current ecology textbooks, more than half a century after he did his pioneering work.

Gause performed two main sorts of experiment, representing what are often thought of as the two main types of ecological interactions between species generally – predator-prey (or vertical) interactions and competitive (or horizontal) interactions, where different species jointly utilize a limited supply of resource and/or

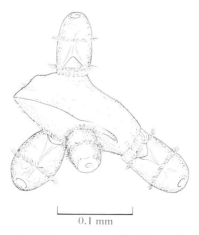

0.1 mm

FIGURE 7   A *Paramecium* being consumed by four *Didinium* predators. Both prey and predators are protozoans.

produce substances (toxins, waste products and so on) which have a negative effect on each other's populations. In both kinds of experiment Gause adopted a multi-generation approach. In other words, he placed predator and prey, or both competitors, together in some sort of experimental vessel (usually a glass tube or flask) containing a nutrient medium, and simply let the populations concerned get on with it. Interference by the experimenter is kept to a minimum in such a regime, and is usually limited to periodic replenishment of the medium coupled with periodic counts of the number of organisms present. The end-result of an experiment of this kind is a simple graph of population numbers of both (or all) of the species involved against time, usually extending over many generations. This is sufficient to observe whether the system ends up reaching some sort of stable equilibrium, or whether it exhibits a trend towards the extinction of one or both interactants.

Tiny Predators and Tiny Prey

Let's deal with the predator–prey experiments first. These involved one kind of protozoan (a species of *Didinium*) consuming another kind of protozoan (a species of *Paramecium*). That is, *Didinium* is the predator and *Paramecium* its prey, as shown in Figure 7. What happens when populations of both species are placed together in an experimental microcosm? There's no single answer to this question, because the result is dependent on the structure of the environment in which the experiment is conducted; but essentially there are two possible outcomes, as follows.

When the environment consists solely of a homogeneous suspension of nutrient medium, *Didinium* is rapidly able to find and consume all the *Paramecium*. The result is that we get extinction of the prey, followed very rapidly by extinction of the predator – that is, the system 'implodes' and effectively disappears. This is instability in its most extreme form.

By changing the structure of the environment slightly, it's possible to alter the outcome of the predator–prey interaction completely. However, it appears that it is *not* possible to stabilize it. Rather, one form of instability can be traded for another. What

Gause did was to add an oatmeal sediment, so that there was an overall habitat consisting of two microhabitats – the nutrient suspension at the top of the tube, and the solid sediment at the bottom. What happened in experiments conducted in this kind of system was that many *Paramecium* entered the sediment, while *Didinium* did not. The predators consumed all the prey that remained in the nutrient suspension, and then, finding no more food, declined in numbers until they became extinct. The remaining *Paramecium* multiplied and eventually re-populated the nutrient suspension as well as the sediment. Thus the predator–prey interaction was again unstable, in that the two types of organism were unable to coexist. However, instead of a 'double extinction', we have only a single one.

It is clear, then, that in these simple environments the typical outcome of the predator-prey interactions studied was instability and extinction – either of the predator alone, or the predator and prey together. Gause did, in a further experiment, manage to manufacture a sort of 'cyclic stability' wherein prey and predator coexisted for a much longer period, but this last experiment was of a very artificial nature. The 'stability' was only achieved by mimicking immigration – that is, by adding individuals to the system periodically. But of course most unstable systems can be made 'stable' by this means, so the result is hardly an informative one.

The impact of Gause's results on predator–prey systems comes across most forcibly if we contrast them with comparable systems observed in nature. If you opened the curtains one morning and discovered all the worm-eating birds in your garden dead or dying, and on digging up the lawn found there to be not a worm in sight you would – I imagine – be extremely surprised. Natural or semi-natural predator–prey systems do not tend to 'implode' as Gause's laboratory counterparts did. Observations of natural 'implosions' are so rare that we tend to take it for granted that the natural systems are automatically stable. Yet if asked to pinpoint precisely what it is that causes our gardens to behave differently from Gause's microcosms, no-one could give a satisfactory answer. We are, however, beginning to have some clues about the answer – more on this later.

Competition between Species

So much for predation. Let's now turn our attention to Gause's other main batch of experiments, on interspecific competition. In these, Gause used two different species of *Paramecium*, and although the precise pair of species differed from one series of experiments to another, I'll concentrate here on the results obtained from systems involving *P. aurelia* and *P. caudatum*. These species were introduced, together, into vessels containing nutrient medium as before; but this time there were no *Didinium* predators and they competed for the limited supply of nutrient material present, and consequently had a suppressive effect on each other's populations. (They may also have had mutually suppressive effects because of their excretion of waste products, but this aspect of the experiments has received relatively little attention.)

As with the predator–prey experiments, we ask: what happens when two populations are placed in the experimental microcosms and left to interact? In the competition experiments, of course, neither species directly consumes the other. Rather, their interaction is indirect, and operates through removal of the jointly used resource (a bacterial suspension). However, the possible outcomes are broadly similar to those that apply to the predator–prey systems: the systems may be stable or unstable. The main difference is that double extinctions ('implosions') are impossible in competitive systems – here, instability always takes the form of single extinctions.

In all Gause's early experiments with *P. aurelia* and *P. caudatum* the outcome was the same: the former species 'competitively excluded' the latter, which went extinct. After extinction, the system was simply a monoculture of *P. aurelia*. It seems that in some of Gause's later work (done after the publication of *The Struggle for Existence*), sets of conditions may have been found which 'flipped over' the result, and caused *P. aurelia* to go extinct – though whether this is the case is not entirely clear, because these later experiments are not as well documented as the earlier ones. What *is* clear, however, is that Gause did not find any set of conditions under which these two species of *Paramecium* could coexist. Again, instability ruled.

It's more difficult, in the case of the competition experiments, to make a striking contrast with nature. This isn't because competitive interactions in nature are predominantly unstable (which they almost certainly aren't), but rather because competition is a more elusive kind of interaction, in natural systems, than predation. It's easy to go into a garden or park and observe an 'act of predation' in which one organism consumes another. Given a day to go out and observe such an act, you would almost certainly be successful. However, a day to observe an 'act of competition' in nature would put your powers of observation to a rather more severe test.

It's almost certainly true that, in a lawn (or similar natural 'grassland' system), different species of plants are in competition with each other. The problem for the observer is that experimental manipulation is required to reveal, in a conclusive way, that competition is indeed taking place. If removal of one species (e.g. with a selective herbicide) results in the spread of another, and the reaction is reciprocal to some degree, then the plants were competing. But no amount of observation on its own will reveal the interaction.

Given this problem of the elusiveness of competition, you'll have to, for the moment anyhow, take my word for it that competition is a widespread interaction in natural communities (though precisely *how* widespread is a matter of some current debate), and that the competing species typically settle down to a state of stable coexistence. So again, the 'typical' laboratory system and its 'typical' field equivalent differ in that the latter is stable while the former is not. And again, the reason why this is so isn't clear; though as with predation we have made some headway at understanding the difference, and therefore at identifying the causes of natural stability.

Dissecting Ecological Complexity

I've often heard it said – by seasoned ecologists who should know better, among others – that the reason why experimental ecological microcosms are typically unstable is that they lack the 'complexity' inherent in natural communities. Now this statement may well be true (and we'll examine it further in Chapter 10), but I doubt whether it conveys any real information. If you asked a television engineer how it was that that glass-fronted box to which you devote so much of your precious time was able to convey motion pictures and sound from hundreds of miles away, he could reply that it was because of the 'complexity' of the electronics inside. Why is it that such an engineer would be regarded as an idiot, while some ecologists can get away with statements of equal vacuousness and be treated as gurus?

What we need to do is to 'dissect' the complexity. We need to break it down into its components and ask which of them are important in stabilizing ecological interactions and which aren't. You can think of this as an enormous programme of experiments in which different components of the complexity are added in, one at a time, to Gause's microcosms. Perhaps, on adding the 43rd such component, the system would 'flip' from an unstable mode to a stable one. If so, you might try adding the components in a different order to see if it's just one of them, or some combination of them, that achieves the stabilizing effect.

Of course, the direct approach to analysing ecological complexity cannot always be taken. We cannot really add in more and more aspects of nature to one of Gause's microcosms until we have a mini-rainforest in a tube. But most meaningful ecological research on the stability question can be seen as indirect approaches to the same goal. These approaches are very varied, and we'll glimpse many of them in the following chapters. They include everything from doing the kind of experiment I've just described in a 'computer microcosm' rather than a laboratory one to the 'opposite' approach of removing components from natural systems and observing the effects of such removals on the stability of the remainder of the system. In all cases, the thrust is the same – analysis of ecological complexity to reveal its 'effective' components.

Once we acknowledge the rationality of this approach, comments about ecological experiments done in the laboratory being 'un-natural' are readily exposed as misguided. It is their very un-naturalness that allows us to make the contrast from which we begin our analysis. The situation is somewhat akin to that in genetics. If you want to find out what a gene does, an obvious approach is to compare the 'wild type' with the mutant phenotypes that result when the gene concerned is defective. Likewise, if ecologists want to know the effect of a particular environmental factor (such as 'refuges' from predators as in Gause's experiments), they need to compare systems having and lacking that factor. This is merely an ecological application of the conventional 'reductionist' approach which has delivered so much understanding in other branches of science; and there's no reason to believe that it should be less informative when it comes to understanding processes like competition and predation than it is in relation to understanding the functioning of genes.

However, this is not quite the end of the story. While ecological processes such as competition and predation may indeed be profitably analysed by the reductionist experimental method, it can be argued that the resulting information is only relevant to the functioning of those interactions when they occur 'in isolation'. It may be, for example, that refuges from predators will have one sort of effect in a 2-species system and quite another effect (or no effect at all) in a 100-species system such as might be found (in the form of a complex food web) in a natural community.

Since this sort of worry is insoluble, we come back to where this chapter started – voting with our feet on the philosophical matter of whether to be a reductionist or a holist. In this book I adopt a primarily reductionist approach for two main reasons. First, because I happen to believe that factors acting to stabilize ecological interactions in simple systems will have a qualitatively similar (albeit quantitatively different) effect in more complex ones. Second, because I remain to be convinced how it is currently possible to proceed beyond a very superficial understanding of ecological systems using any non-reductionist approach. I suspect that in the end we will only fully comprehend natural communities by a combination of holistic and reductionist tactics; but at the present embryonic stage in our assault on the balance of nature, I believe that more progress can be made via the reductionist route. Ecology being the healthy flux of contradictory views that it is, some ecologists will beg to differ.

6

METHODS FOR MISSION IMPOSSIBLE

No ecologist would ever pretend that our quest for the general principles underlying the balance of nature is an easy one: it's either very difficult or impossible. I happen to believe the former, or I wouldn't be writing this book. However, 'Mission Very Difficult' doesn't have much of a ring to it, so I chose to forego accuracy for impact and entitled this chapter according to the pessimist's view.

We're now approaching the 'core' of the book (roughly Chapters 7 to 14), which is about the results of ecologists' endeavours so far, both in terms of selected case-studies of particular systems and in terms of the general ecological principles that we think we can dimly see as a result of these, and other, studies. Before going on to this 'core', though, we need to take a brief look at some of the methods ecologists use. A detailed description of all ecological methods would be out of place here, so I'll be extremely selective and outline only those methods which are absolutely essential to understanding the results of these ecological endeavours.

In fact, I'll concentrate on just two things. The first is the general business of 'sampling' – that is, the art of learning about something very large (such as the boreal forest) by studying only a very small part of it (such as a few 50-metre-square sample sites), yet being reasonably confident that our findings are not misleading. The second is the business of estimating the size of a natural population of animal, plant or microbe. This is really a particular application of the sampling ethos, but it's sufficiently important to warrant separate treatment. Speculations about whether a particular community is 'stable', whether a particular species is 'endangered' and so on are ultimately based on estimates (or sometimes, unfortunately, guesstimates) of the number of organisms present in particular places at particular times; so it's important to know how such measurements are made and how accurate they are. It's no problem to count the number of cows in a field, but the number of whales in the Arctic Ocean or the number of flies in a forest is not quite so easy to ascertain.

Sampling

And so to sampling. It's important to realize that almost every ecological 'fact' that we claim to know is based not on a study of the whole system in which we are ultimately interested, but rather on a study of small samples of it. Sampling is ubiquitous in ecology. If our sampling methods are somehow biased, all our conclusions may be wrong. Nothing is more imperative for a practising ecologist than an understanding of how to make samples 'representative'.

One of the most forceful ways of emphasizing this message is to illustrate what can happen when the potential dangers of biased sampling are ignored. Such illustrations can be drawn from fields quite different from ecology, since many other fields of study also have sampling as a central methodology. A well-known use of sampling techniques in the Western world is the political opinion poll. Here, interest ultimately lies in how 'the electorate' will behave. Since this is a body of around 40 million people in the UK, and in excess of 150 million in the USA, it clearly cannot be studied in its entirety. Instead, a sample is taken, and it is often quite a small one, given the size of the electorate. Recent British opinion polls (which to their credit usually publish the sample sizes involved) are often based on about 1000 people – less than a tenth of 1 per cent of the total. Yet, because we now know quite a lot about sampling theory, they are often reasonably accurate. In contrast, an early opinion poll in the USA conducted by telephone at a time when few people had them produced a hopelessly biased result – biased, of course, in favour of the political views of the richer social strata, which then had by far the highest proportion of phones.

While this is a particularly clear example of the dangers inherent in sampling, I don't mean to give the impression that only cases of spectacular bungling will produce erroneous results. Standing on a street corner and asking passers-by how they will vote is clearly more satisfactory than polling by phone, since there is no inherent bias in favour of phone owners. Yet for all the apparent randomness of the street-corner questioner, such a poll conducted at 8 a.m. might give quite a different result from an otherwise identical one conducted at 9 a.m., for the simple reason that there are statistical differences in the going-to-work times of members of different social strata as well as in their voting tendencies.

Returning to ecology, we find that the dangers facing the would-be sampler are greater than in the case of students of human behaviour. This is because we know so much less about the habits of the other inhabitants of our planet than we do about our own. We don't usually know, for example, the extent to which apparently identical animals possessing different sets of genes behave differently. It's therefore less easy, when sampling an animal population, to anticipate potential biases and so to try to avoid them. Nevertheless, it's easy to imagine situations in which a particular sampling procedure would lead to the wrong conclusion.

Suppose, for example, that you are given the job of determining the abundance of a particular species of moth in a very large area of forest. You might designate a particular site, say one hectare (a hundredth of a square kilometre) in area, as your study site, and then use a sampling procedure to estimate the population size. (We'll see precisely how this can be done shortly.) Even if your estimate is exactly correct for the site you have chosen to study – an infrequent state of affairs to say the least – there are considerable problems in extrapolating this estimate from the small study site to the overall area of interest, which may stretch to thousands of square kilometres. One of the main problems in doing so is that species' distributions often tend to be very patchy. That is, there are places where density is very high, places where it is very low, and places where it is intermediate – with the variation often occurring for no apparent reason. Given a patchy distribution, and only a single study site, how do we know if we have sampled a high-, low-, or average-density patch? Clearly, the answer is that we don't, unless several other study sites have been sampled as well.

What applies in space applies also in time. Populations fluctuate – sometimes quite markedly – and an estimate of abundance made in a single year may therefore be quite misleading. We all know of years of ladybird plagues, years of almost no wasps, and so on. Of course, if estimating the population size of one of these familiar beasts, we would *know* if we were operating in a particularly atypical year. But much ecological work involves organisms of which we are more radically ignorant. In such cases, the solution to the problem of temporal variation is the same as in the case of its spatial equivalent – take many samples, not just one. Exactly how many should be taken depends on a variety of factors, and this isn't something that we need to go into here; but the message of multiple rather than single samples comes across loud and clear. And it is not a message that is unique to estimates of population size – it applies equally to estimates of other ecological variables such as the species richness of a community.

Given that we need to take several samples in order to obtain any meaningful ecological information, how should these be arranged in space or time? That is, what *pattern* of sampling is optimal? The answer is that it depends on exactly what you're trying to find out. If you're simply trying to get a picture of the 'typical' abundance of an organism in a particular habitat type, or, for that matter, trying to assess the 'typical' species-richness of the habitat, a *random* scattering of samples is best. However, more complicated ecological questions demand more complicated arrangements of samples. In a study in which the effect of a tree canopy on the relative abundances of two species of land-snails living in the ground flora is being investigated, *paired* samples of neighbouring wooded and non-wooded areas are optimal. A search for an altitudinal trend in species richness would probably be best served by some sort of *transect*; that is, a more-or-less linear array of sampling sites, in this case ascending a hillside. And finally, an investigation of the population growth of a newly-arrived migrant species in a particular area would probably be based on a series of samples spaced *regularly*, rather than randomly, in time: for example, annual censuses at the peak of each breeding season. This enables the investigator to look at the underlying trend in population size denuded of the expected seasonal oscillations – in much the same way that economic analysts are more interested, for example, in how this January's unemployment figures compare with last January's than with the August in between. (The latter sort of comparison in itself is meaningless since it compounds seasonal and long-term effects.)

Estimation of Population Size

Let's now restrict our attention to the use of sampling procedures in estimating population sizes. I'll assume that we wish to estimate the 'absolute' population size (and so produce an estimate of the form "1257 moths") rather than merely the 'relative' population size ("this area has more moths than that one"). There are two main ways of estimating absolute population sizes, and the difference relates to the mobility of the organisms with which we're dealing. One method, which I'll call 'count-and-multiply', is primarily used with sedentary organisms, and therefore mainly with plants. The other method – 'mark-recapture' – is used when the organisms concerned are mobile, and is therefore a predominantly zoological technique. I'll describe these two methods in turn.

The rationale behind the 'count-and-multiply' method is as simple as its application in most practical situations is complex. (Actually, the same could be said of mark-recapture.) Let's consider first its application to a Forestry Commission conifer plantation, which, because of the regular distribution of the trees, is much easier to deal with than most natural habitats. Suppose that, due to a loss of records, it is no longer known how many trees there are in a plantation measuring three by two kilometres, and you are employed as a 'consultant ecologist' to resurrect the lost information. You could, of course, walk the length and breadth of the forest going "one, two, three …" and putting a dab of white paint on each tree counted to ensure no omissions or double-countings. However, this method would be ludicrously laborious. A much quicker, and almost equally accurate, method would be to count the number of trees in a small sub-area – say a one-hectare square (100×100m), and then multiply up by the appropriate factor (600 in this case) to obtain the total.

Why is this method not so easy to apply to most *natural* populations of plants, which are just as immobile as the trees in the forestry plantation? There are four answers to this question. Firstly, and perhaps most trivially, there is the physical layout of the habitat unit. The Forestry Commission is notorious for its tendency to plant rectangles. From an aesthetic point of view, this is a criminal destruction of the landscape, but from a calculational viewpoint it simplifies things. The area of a natural forest with a ragged periphery is not so easy to determine. However, an aerial photograph and a digitizer would soon deal with this problem, so we shouldn't get too concerned about it. Second, many forestry plantations are monocultures (for example of Norway Spruce), so there is no species problem. A species of grass in a natural grassland is another matter entirely. But again, the problem can be solved – in this case with a good taxonomic 'key'.

The other two problems are more serious. Third on my list is the difficulty, in some plant species, of determining what constitutes an individual organism. Animal ecologists usually take recognition of an individual for granted and immediately focus their attention at the population level. If you wish to determine the number of red deer on a Scottish island, you don't lose too much sleep over what constitutes 'a deer'. However, grasses and many other plants present a different picture entirely, with underground connections linking apparently separate above-ground 'individuals' whose genetic constitution is identical. Indeed, a whole field of grass might be a single 'clone'.

While this is a very serious problem to ecologists working with certain groups of plants, it can be avoided by those working on other groups. The final problem, however, is ubiquitous and cannot be got around by a shrewd choice of species. This is the problem of patchy distributions, which we encountered a little earlier. Unfortunately (again from a calculational, rather than aesthetic, viewpoint) most species are extremely irregular in their distribution. Trees in a natural forest, for example, are not laid out in neat rows like their plantation cousins. Counting a small sub-area and multiplying up is, therefore, fraught with problems.

As I mentioned when we first encountered this difficulty, a first step in its resolution is to take multiple samples. What do we do then to estimate the population size? The simplest solution is to average the sub-area results and then multiply up. The extent to which this works depends on just how patchy the distribution is and on how many sub-areas have been counted. Obviously, the more sub-areas the better, but equally if we take too many then the workload is not much less than counting the entire population. Each situation must be assessed individually to determine the optimal strategy.

The Problem of Mobility

While the four problems just discussed can be severe, they are as nothing compared with the problems that arise from movement of the organisms whose population size is to be estimated. Consider, for example, a population of a species of woodland bird. Suppose you are given the job of estimating the size of such a population in a particular patch of natural woodland which, like the forestry plantation we considered earlier, has an area of around six square kilometres.

If you tried to estimate the population size by a 'count-and-multiply' technique your estimate would probably turn out to be extremely inaccurate – and the reason for this is obvious. As you walk through your study area (say 100m×100m as before), you might count a curious bird (who follows you), several times, while a shy one (who retreats from you) would not be registered in your count at all. In the end, your 'count', even of the limited 100×100m study area alone, would be meaningless; and the large error it will inevitably contain is then multiplied by 600 in the production of an overall population estimate for the whole wood.

There are two ways around this problem. The first is somehow to 'freeze' the organisms concerned, so that they are effectively immobile. Of course, you cannot literally stop a bird in mid-flight or a lion in mid-chase. But what you can do is to arrange things so that you can disregard the mobility. Suppose, for example, that you want to determine the sheep population of a particular sector of the Northumberland National Park. You could wait for appropriate weather conditions and take a high-resolution aerial photograph. Effectively, you have 'frozen' the sheep. Estimating their population size is then relatively easy.

Of course, this technique is only useful when the organisms you are dealing with are reasonably large and when their habitat is of the 'open' (e.g. moorland) type rather than the 'wooded' type. Aerial photography of the woodland bird with which this discussion started might give a population estimate of zero! In such cases, it's sometimes possible to count static products of the animals rather than the (mobile) animals themselves. In the case of the woodland bird, an obvious choice would be the number of nests in the nesting season. This would also have the advantage of ensuring that you are estimating the resident population and excluding any individuals that are 'just passing through'.

The alternative to the 'effective immobility' approach is to *accept* the mobility – even to make use of it – by employing a completely different technique of population size estimation: mark-recapture. Again, as with all the versions of the count-and-multiply technique, the rationale behind the methodology is very simple, but its application to particular natural habitats often is not.

In its implest version, the mark-recapture technique consists of the following. To provide a mental picture, let's suppose that we are estimating the population size of brown trout in a small lake. First, a sample of fish is taken and the individuals marked. The method of capturing the animals might in this case involve the use of a large net; and the marking procedure might be to affix a small 'tag' to one of the fins. Clearly, the actual capturing and marking techniques are very species-dependent: if we were estimating the population size of a water-snail in the same lake, for example, a dab of paint on the shell would be a more appropriate marking method.

The second step in the procedure is to release the marked

animals back into their habitat – the lake in this case – and to leave the habitat undisturbed for long enough to ensure that the marked animals have mixed back in with the rest of their population. After this wait, the third and final step is to take another sample (which will probably be roughly similar in size to the first, though this isn't necessary), and to count the number of individuals in the second sample which are marked. These are the 'recaptures' – i.e. the fish that were captured on the first occasion and have now been captured again.

With a few provisos that we'll look at in a moment, the estimation of population size is now quite simple. Assuming that our second catch is a representative sample of the total population, the proportion of marked fish in that sample will reflect the proportion of marked fish in the lake as a whole. Since we know three out of four of the necessary values – i.e. there is only one unknown, the population size – the missing value can easily be deduced. For instance, suppose our first sample contains 50 fish, while the second contains 60, of which 10 are marked. If this sample is indeed representative, then about a sixth of the total population is marked. So our question becomes: what number is 50 a sixth of? The answer, of course, is 300, which is then our estimate of the population size for brown trout in that particular lake.

Let's now look briefly at the 'provisos'. There are basically three of these, in addition to the need for intermixing of marked and unmarked individuals, which we've already noted. First, the mark must be durable. If tags fall off or paint-marks dissolve, our estimate of population size will be erroneous. Second, if the marks significantly affect the animal – for example, making it more conspicuous and therefore in some cases more likely to appear in the second sample than an unmarked animal – then again we'll get an erroneous estimate. Finally, we've assumed that the same animals are present in the population throughout the mark-recapture process. If this is not so – that is, if animals are born, die, immigrate and emigrate – then again the estimate is adversely affected.

Some of these potential problems may be prevented by additional practical work. For example, if you want to know whether a dab of enamel paint will come off a pond-snail's shell in a three-week period, you can easily set up a laboratory aquarium population to investigate this. Other problems – such as migration or births/deaths – can be dealt with by the use of a more sophisticated mark-recapture programme involving more than two sampling events. Such programmes (which lead to much more complex calculations) are well explained in a book called *Estimating Animal Abundance* by the British ecologist Mike Begon.[9]

If we arrange things such that all the provisos are satisfied, will our estimate of population size then be reliable? The answer is: to some extent. Just *how* reliable it is will depend, among other things, on how big the samples are. Like all other sampling procedures, mark-recapture is subject to the inevitable 'sampling variation', and the smaller the sample sizes, the worse this problem gets. As usual, a sensible response to the problem is to replicate our efforts. If two independently conducted mark-recapture programmes carried out on our fish population at about the same time produce estimates of 300 and 314, we begin to feel that we are getting close to the truth. However, if the estimates are 300 and 15,000, then there is clearly a long way to go. This shouldn't be taken to imply that consistent results are *necessarily* also accurate ones, since it is possible that repeats of a sampling programme contain similar biases to each other. We can minimize this problem, however, by making the two (or more) mark-recapture procedures different from each other in many respects. If they are conducted by different ecologists under different weather conditions and using a different method of marking, yet they produce consistent results, then we can be more confident of their accuracy than otherwise.

Making Use of the Estimates

So much for the estimation of population size. Assuming (as I will from now on) that our estimates are at least tolerably accurate, what useful information do they convey? In some cases, the answer to this question is obvious. For example, if the ecosphere contains only a single population of a particular species, the extent to which that species should be regarded as 'endangered' is very much dependent on what number of individuals make up its single population: if only ten, it may soon be lost forever; if ten million, it may not only be out of immediate danger but may be on the point of a range expansion, after which it will be even less endangered.

In other cases, an estimate of population size conveys rather less direct information. Let's consider two aspects of this. First, many ecological processes – such as competition, which we'll look at in Chapter 8 – depend not so much on population *size*, but rather on population *density*. If our population of 300 brown trout is located in a small pond, we might expect there to be very intense competition for food or space. If, on the other hand, they occupy a lake of several hundred square kilometres, then there may well be no competition at all.

Sometimes, the conversion of population sizes into population densities is simple: we just divide the number of organisms present by the area or volume of the habitat. For fish, the habitat is three-dimensional, so we use volume; and it's not too difficult to determine the approximate volume of a body of water such as a lake. For sheep or trees, the habitat is effectively two-dimensional, so we divide by area, whose estimation is again reasonably straightforward.

On other occasions, the conversion to density is more problematic. The density of a leaf scale insect, such as greenhouse whitefly, is not optimally described either in terms of numbers per unit land area or numbers per unit plant-habitat 'volume'. Probably, the best form of density estimate for such a beast is numbers per unit leaf area, the latter being very difficult to determine in many cases. Also, while leaf area may be the best variable to use in connection with the actual 'scale' (the pupal, or resting stage, which is actually affixed to the leaf), it may not be the best variable to use in relation to *adult* density. Indeed, the whole business of population size and density estimation in organisms with multiple lifestages is problematic, the more so the more different the ecologies of the different stages. And since the majority of the world's species of living organism are insects, where multiple-stage life-histories are the norm, this problem is hardly a **minor** one.

There is a second aspect of the indirectness of population size information. We encounter this when our aim is to look at population *dynamics*, that is, how populations change over time. A single estimate can tell us whether, *at that instant*, the population size is large, small or middling. But whether it is growing, declining, or remaining approximately constant is another matter. To assess this, we need a string of measurements in time — probably regularly spaced (e.g. one a year), as suggested earlier. The three chapters that follow examine, in the context of various ecological processes such as competition and predation, just such time series of population dynamics data. As we'll see, the analysis of this kind of data sheds much light on our goal, which, as ever, is an understanding of the balance of nature.

7

Why is Extinction so Rare?

Allow me to introduce you to the Mongolian blue-tailed thumper. This is a small squirrel-like creature, mostly chestnut-brown in colour, but with a bright blue stripe running the full length of its flattened tail. Thumpers get their name from the tendency of the male to attract females by thumping the ground with its tail during the mating season – the harder the thump, the greater the male's attractiveness. I want to focus our attention on a particular population of thumpers, living in a forest in the Kerulen Valley. For reasons I'll come to shortly, I have a very detailed data-set for this particular population, and you can see this displayed in Figure 8.

The history of this population consists of five distinct phases or events. First, the population came into being through immigration of a handful of individuals from a previously-established population in a neighbouring valley. Second, there was a phase of considerable population growth wherein, although there were a few short-term declines, the underlying trend was consistently upwards. The third stage was a prolonged period – more than a hundred years – characterized by continual fluctuation in population size (populations rarely if ever remain precisely constant) but no long-term upward or downward trend. Following this, the fourth phase was one of rapid decline, mirroring the earlier growth-phase. Finally, in 1985, the population became extinct.

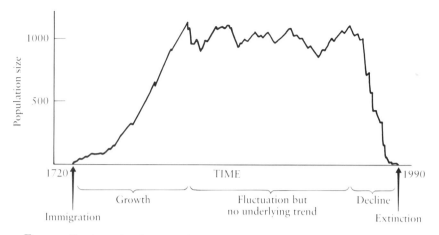

FIGURE 8   Complete history of population size for the colony of 'Mongolian
blue-tailed thumpers' living in the Kerulen Valley.

As it happens, virtually *all* populations of all species display this
general sort of pattern. All populations certainly have limited
lifespans bracketed by immigrations and extinctions. While some
propagules of migrants invading new areas may simply go extinct
because the habitat is unsuitable, such a group of organisms would
hardly be said to constitute a population. Assuming the habitat *is*
suitable, then a period of population growth will take place. Its
mirror-image, the decline phase, is also almost certain to occur,
since extinctions are rarely instant drops to zero from a high
starting point, though they do vary greatly in their rapidity. That
leaves us with the middle phase, where population numbers
exhibit little if any net change over a prolonged period of time.
Again, this is near-universal, since population growth giving way
immediately to decline and extinction is rare; but how long the
middle phase is – whether a handful of generations or a matter of
hundreds or thousands of generations – varies greatly from one
system to another, as do the magnitude and pattern of the
fluctuations.

One additional complication that some species exhibit is a
regular cyclic oscillation in population size, with peaks or troughs

separated by a fairly constant period of time – often around 3–4 years or 10–11 years, though the reason for the prevalence of these particular periodicities is unknown. Cyclic populations are not fundamentally different to non-cyclic ones: both exhibit the five phases illustrated in Figure 8. In cyclic ones, we simply have an additional feature to explain. (Note that *most* populations exhibit seasonal cycles, but these are invisible in annual-census data-sets). I should emphasize that cyclic populations are very much the exception rather than the rule – a fact which may be obscured by the disproportionately large amount of attention that ecologists have given them. The best-known examples are some rodents (3–4 year cycles) and some larger mammals such as the Canadian Lynx (10–11 year cycles). But most mammals, like most other creatures, appear to have essentially non-cyclic populations.

It's probably time to admit what you may well have already suspected: that while the Kerulen Valley in Mongolia is a real part of the ecosphere, the 'blue-tailed thumper' is an entirely fictional inhabitant. Ecology as a quantitative science has a short history (less than a century) and consequently there have been no studies in which population sizes have been monitored by mark-recapture or count-and-multiply techniques over periods of hundreds of years. It's much more common to find data-sets that essentially comprise a time-slice out of a population's history. For example, many references could be made to ecological research articles which display population dynamics data extending through a period of ten or twenty years somewhere in the 'middle phase' of the populations concerned – comparable, for example, to the thumper's population trajectory from 1850 to 1870. Such data exist for many different species; an example is given in Figure 9.[10] Complete population histories (except for very short-lived populations) are not yet available, which is why I decided to invent one to give you the full picture.

Having seen this picture, which of its features are obvious, and which, on the other hand, urge us to seek an explanation for them? I'd argue that most (four out of five, in fact) have what you might call commonsense explanations and do not seriously tax the ecological mind. This applies particularly, I think, to the initial and final events of population history, that is, the population's

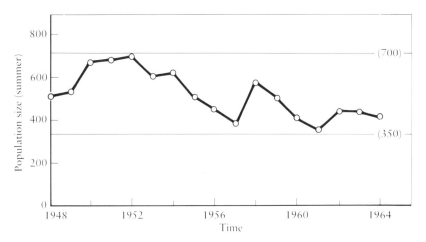

FIGURE 9   Slice of population history for pied flycatchers in the Forest of Dean, Gloucestershire, England. (Fine horizontal lines show approximate limits of the fluctuation pattern.)

origin, through immigration, and its eventual demise, through extinction. When separated off from the growth and decline phases with which they are respectively associated, this leaves the actual immigration and extinction events as very small-scale phenomena with large random components. Precisely how and when the first few migrants arrived in an area is usually unascertainable in individual cases; and the fact that migration occurs in general hardly requires any deep explanation, since most kinds of organism have some mobile stage in the lifecycle – even ones that we normally think of as static, including higher plants (pollen), ferns (spores) and barnacles (planktonic larvae). Equally, the causes of death of the last few remaining members of a population that is going extinct may be no different from the causes of death during an earlier phase of the population's history; so even if we were able to determine why the last few individuals died – thereby causing extinction of the population – the information might not be particularly enlightening.

Growth and decline are both longer-term processes than initial immigrations and final extinctions, the latter events being, arguably, instantaneous. Yet they still present us with little in the way of a conceptual challenge. All organisms can potentially reproduce, and all organisms die, and if the net effect of these is positive, we have growth, if negative, decline – assuming no complications to the picture due to further migrations after the initial one. Habitats change over time, sometimes slowly, sometimes abruptly (the latter often associated with human activities), and the idea that an initially favourable locality (for a particular species), will one day be unfavourable is perfectly straightforward.

All this is not to say that immigration, growth, decline and extinction have nothing to offer the ecologist. Particular cases may be interesting for various reasons. For example, ecologists have invested considerable effort (and continue to do so) in attempting to understand whether the decline of red squirrels in Britain is linked (via competition) with the spread of the immigrant North American grey squirrel. Also, *comparative* information on colonization and extinction rates has been instrumental in the development of the theory of island biogeography, as we noted in an earlier chapter. However, specific cases and comparisons aside, the general *idea* that, at some stage in their history, populations grow and decline presents no great conceptual challenge.

The same cannot be said when we turn to the remaining phase of a population's history – the middle phase during which, despite numerous short-term fluctuations, the population exhibits no net growth or decline. A population of 1000 could, three generations on, consist of a million individuals (on the assumption of a moderate ten offspring each), or consist of a single individual on the verge of extinction (on the assumption of 90 per cent mortality, which is again moderate for many kinds of organism). In the face of such massive potential for growth and decline, how can the two so precisely balance each other out that, over a hundred generations (say), the population size exhibits virtually no net change? Does such a balance not seem almost infinitely improbable?

The answer to the second question is of course 'yes', unless there is some active mechanism working to produce a balance. For example, if a population is characterized by a multiplicative ability of 10× per generation *and* 90 per cent mortality per generation, then it will indeed remain the same size. If, in a population of 1000 trees, each tree has ten seedlings but seedling mortality is 90 per cent, each new age-class of trees will, on maturity, be 1000 in number. But if the mortality rate changes to 89 or 91 per cent, or the average number of seedlings per tree alters to nine or eleven then, in a hundred generations, we have massive potential for long-term growth or decline. In natural systems then, there are two great improbabilities threatening the long-term constancy of population size. First, it is very improbable that birth and death rates should ever be *precisely* equal. Second, even if they are equal in generation 42 (say), it is very improbable, given the vagaries of nature, that they would still be equal a couple of generations later. Since it's hardly conceivable that populations keep managing to somehow overcome these immense improbabilities, and ensure that in each generation birth and death rates are precisely equal (in which case, incidentally, there would be no fluctuations), we return to the idea that some active *balancing mechanism* must be at work, and the nature of this mechanism now becomes the focus of attention.

An Ecological Thermostat

If I said that what prevents population size from increasing or declining too much during the middle phase of its history is a thermostat, you might well regard that as a rather weird proposition. Yet in a sense (and providing you'll permit a fairly broad definition of a thermostat) this is precisely what happens. Consider for a moment a *real* thermostat, such as you may have in your living room to control the ambient temperature which the central heating system maintains. The way such a device works is not by keeping the temperature absolutely constant (which is impossible), but rather by *constraining the fluctuations* in temperature into a fairly narrow band. If the temperature is proceeding towards the upper limit of this band, the boiler is switched off, so the temperature begins to drop. But if it continues to drop to near the bottom of the target band, then the boiler is switched on again, so the temperature goes back up, and so on indefinitely.

When this kind of control system is encountered in organisms – and I do mean *in* organisms, and so in the physiological rather than the ecological realm – it's generally referred to as homeostasis. Again, the variable being controlled may be temperature. For example, our own body temperatures (and those of other mammals) are maintained at around 37°C (or 98°F) by a homeostatic mechanism. If we're hot, blood is diverted to the surface (red face, 'swollen' veins) which results in heat loss to the (normally cooler) atmosphere. If we're cold, blood is diverted *away* from these peripheral areas so that heat loss is minimized. Physiological homeostasis also produces a state of constrained fluctuation in lots of other variables as well as temperature, such as the concentrations of sugar and oxygen in the blood.

What's called thermostatic control by heating engineers and homeostasis by physiologists is referred to by population ecologists as 'density dependence'. What, then, *is* density dependence apart from the fact that, by analogy, it's a mechanism that produces a pattern of constrained fluctuation in population size?

Since density dependence is a complex subject, let's begin to discuss it in a relatively simple system with the following features. First, we focus on a particular population: it could be blue-tailed thumpers in the Kerulen Valley, wildebeest in the Serengeti, pine trees on a wooded island, or whatever – choose whichever mental picture you find most appealing. Next, we make the assumption that while the population will fluctuate in size, and the individuals that comprise it will in some cases be mobile, the total area occupied by the population remains constant, so we can effectively equate changes in number with changes in density. Finally, we assume that our population is 'closed' – that is, immigration and emigration over the study period are negligible, and thus all changes in population size/density are due to births and deaths.

Again to simplify things, let's deal with births and deaths separately, though later we'll consider their combined effect. The overall phenomenon of density dependence is made up of either or both of two components: density-dependent mortality and density-dependent natality, and I'll deal with them in that order.

The phrase 'density-dependent mortality' refers to *rates* of mortality (expressed, for example, as percentages) being positively related to population density. That is, as density goes up, so does the percentage mortality. Notice that this is more than just the *number* of animals or plants that die per unit time being greater when the population is bigger, which in itself is to be expected, and is of little interest. Rather, when mortality is density-dependent the *proportion* of the population that dies becomes greater as population size increases. If such a state of affairs prevails – and we'll shortly look at some of the biological reasons why it often does prevail – then we have a sort of 'brake' on population growth, or, if you prefer, a thermostat.

Density-dependent natality is a similar sort of 'brake' but applied now to the birth rate rather than the death rate. It refers to the natality *rate* (again, note that we're going beyond absolute numbers) being *negatively* associated with population density. That is, as the population size increases, the probability of individuals in it having offspring declines.

It's generally agreed that some ecological factors tend to be associated with density-dependent effects while others do not. For example, competition for food or space is negligible at very low density, but may be acute at high density. Thus competitive effects on natality and mortality rates would be expected to be density dependent. On the other hand, the effect of an unusually harsh winter on a population would be expected to be independent of its size. Whatever the proportion of individuals dying of cold in a low-density population, there's no reason to expect a comparable high-density population to differ. That is, climatic effects are often density-independent.

Of course, any given population has a complex of environmental factors acting simultaneously to determine its natality and mortality rates. Whether these rates *overall* are density-dependent is what we really need to know. With one proviso (which we'll get to shortly), an overall mortality or natality rate will be density-dependent *if at least one of its components is density-dependent*. For example, in a population whose overall mortality rate is composed of competitive and climatic components which are respectively density-dependent and density-independent, the overall rate will be density-dependent. The climatic component can dilute, but cannot negate, the density-dependent effect. This is a bit like saying that if you merge a hillside and a flat plain, you'll always end up with a hillside.

The proviso is simply that none of the components of the overall mortality or natality rate are *inversely* density-dependent: that is, percentage mortality dropping or natality rate rising as density increases. While such a state of affairs cannot be ruled out, it is thought to be relatively rare, since it is difficult to envisage environmental factors which would act in this way.

Obtaining evidence of density-dependent mortality and natality rates in natural populations is not always easy, and the problems encountered when attempting to obtain such evidence are the subject of much current debate among ecologists. However, it's generally agreed that most populations do experience density-dependence. The British ecologist Mark Williamson, reflecting on an earlier era of ecological research in which the commonness of density dependence was less widely accepted, noted that the early debate was now essentially over, and stated that "any population which persists for ten generations or more is likely to show density-dependent effects."[11] Figure 10 illustrates two examples of density dependence in natural populations, one involving natality, the other mortality. In the former case body size is used as an indirect estimate of natality. This is possible because there is a very widespread trend in most invertebrates (and also in many vertebrates and plants) for body size to be positively correlated with number of offspring. Put crudely, big females lay more eggs – or in the snail's case, big hermaphrodites lay more eggs.

I said a little earlier that we would eventually merge natality and mortality rates and look at their net effect which, essentially, is the population growth rate, using 'growth' in the broad sense to include decline (as negative growth). Population growth rates are normally expressed on a per capita basis, so that populations of different sizes can be directly compared. If the per capita growth rate is zero, the population size is constant; if it is positive, the population is expanding; and if negative, declining.

The principle governing the merging of birth and death rates to give a net growth rate is the same as that governing the merging of various sources of mortality to give the overall mortality rate: that is, if at least one of the things being merged is density-dependent,

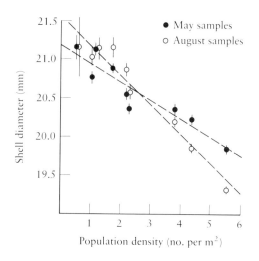

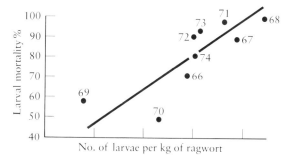

FIGURE 10   Density dependence in natural populations. *Top*: Density-dependent natality (acting through body size) in the snail *Cepaea nemoralis*. *Bottom*: Density-dependent mortality in the moth *Tyria jacobaeae*, which consumes ragwort.

then (with the same proviso as before) the net result after merger is also density-dependent.

Figure 11 shows a plot of a population's per capita growth rate against its density. This is a hypothetical data-set: you can think of it as resulting from some research done on the fictitious 'thumper' with which this chapter started. It clearly shows overall density

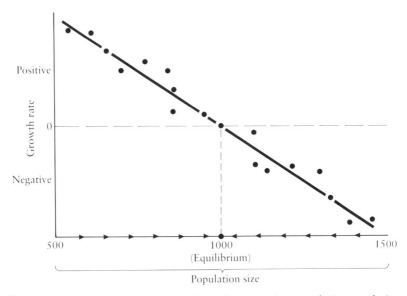

FIGURE 11   Illustration of density dependence causing population regulation.

dependence – that is, a negative relationship between growth rate and density. Like Williamson, I suspect that most populations have density-dependent growth rates, despite the difficulties involved in demonstrating this in some particular cases. The interesting thing, then, is how this phenomenon behaves in a thermostatic way to regulate the population size – that is, to constrain its pattern of fluctuation within a reasonably narrow band around the equilibrium value.

We can use Figure 11 to think our way through this problem. Suppose that at a given moment in time the population size is 500. As can be seen from the graph, at N = 500, the growth rate is positive so the population size increases, as shown by the arrows pointing to the right on the horizontal axis. However, when N = 1000, the growth rate becomes zero and the population size becomes constant. Similarly, if we start with N = 1500, the population declines (leftward arrows) until it reaches 1000, at which point constancy again ensues.

Of course, given the vagaries of nature, we don't ever expect the population size to remain constant for very long. Some 'one-off' event may deplete or boost it dramatically. However, the beauty of the system is that it keeps on compensating for any such event. Whenever the population size is pushed away from its equilibrium, the density-dependent regulatory mechanism keeps bringing it back.

There are, as you might expect, limits to the stabilizing power of any density-dependent system. For example, there may be a 'threshold value' – a minimum viable population size – below which growth rate is negative, perhaps because of the difficulty of finding a mate. Equally, there may be an upper limit to the stability, which when exceeded (e.g. due to large-scale immigration from a neighbouring population) leads to total decimation of the available resource and consequent extinction before resource regeneration can occur. But such effects merely place limitations on the stabilizing power of a density-dependent system; they do not alter the fact that the system is essentially of a stabilizing nature.

Within the limits set by minimum and maximum viable population sizes, will the pattern of density-dependent population growth exhibited by a population be linear, as shown in Figure 11, or will it take some more complex form? There can be no general answer to this question, since the precise pattern prevailing will differ from one population to another. However, the American ecologist D. R. Strong[12] has suggested that it might be quite common for the downward slope to be steep at the start and end, and shallow – even flat – in the middle. That is, growth rate may be relatively unresponsive to minor variations in density in the vicinity of the equilibrium. Indeed, if this were true there might be no single equilibrium value, but rather a broad, rather hazily defined, equilibrium zone. Strong calls this sort of behaviour 'density-vague'.

It's possible that some populations subject to density dependence settle down neither to a neat clean equilibrium value, nor to a narrow band or zone of values, nor even to some sort of cyclical pattern, but rather wander in a very erratic yet deterministic path, never settling down at all (though not going extinct either). This kind of behaviour is referred to as 'chaos'. It isn't yet clear to what extent ecological systems exhibit chaotic behaviour.

Sources of Density Dependence

Given the importance of density dependence as the cause of the constraint in the patterns of fluctuation in population size that we often observe in nature, it is of considerable interest to the ecologist to determine precisely which ecological factors tend to act in a density-dependent way. If we can do this, then we'll have identified the main causes of ecological stability at the population level.

I have already given competition and the climate as examples of density-dependent and density-independent processes respectively. In fact, it's possible to make a more general distinction along the same lines: biotic influences on birth and death rates tend to act in a density-dependent way, while abiotic influences are predominantly density-independent. While such a sweeping generalization must always be treated with some suspicion, there is without doubt a clear difference between biotic and abiotic factors in this respect, even if it's not an absolute difference. So if we want to understand the principal causes of a population's stability, we should examine first and foremost its biotic interactions.

All populations are both consumers of food and food for consumers. Consequently, their major biotic interactions are with their resources – below them in the food web – and with their natural enemies – above them in the web. I'll defer discussion of interactions with natural enemies until Chapter 9, and concentrate, here, on interactions with resources. To some extent, however, these two are the same thing looked at from different directions.

Let's assume for the moment that the population which we choose to study is limited by its food supply. So that we can picture something concrete, let's imagine that we are dealing with a species of snail which (unusually) is dependent on a particular food-plant. A food-limited population of our snail is one which would respond to an increase in the population size of its food-plant with an increase in its own population, but would not increase subsequent to removal of one of its own natural enemies (a bird predator say), because the decrease in predator-induced mortality will be compensated for by an increase in competition. A population of this kind may be stabilized by competition for food, because at high density food is scarce, so mortality is high and natality low, thus leading to decline; while at low density the opposite state of affairs prevails, so the population grows. The three-way link between competition, density dependence and equilibrium population size is clear.

In such a situation, 'competition' refers to the struggle between the individuals that make up the snail population to acquire a portion of the limited resource, that is, the population of the food-plant. This type of competition is only one of several types, and it can be distinguished from the others in various ways. Competition implies mutually inhibitory effects between organisms. Where these effects are achieved by some form of direct attack (e.g. secretion of toxins), we have interference competition. Alternatively, if the inhibitory effects are achieved by sharing a limited supply of resource, then we have exploitative competition – which is what is happening in our snail example. The example also involves intra- rather than inter-specific competition, because the organisms struggling to acquire the limited food supply all belong to the same species population. Finally, our snails are said to compete by 'scramble' rather than 'contest'. This is a rather more subtle distinction. In scramble competition, all the competing individuals get some of the resource with the result that, if density is very high, it is possible to get 100 per cent mortality because no one gets enough food to survive. In contest competition, on the other hand, an item of resource is contested until one individual acquires it and the other 'gives up'; and consequently some individuals always survive and we do not get 100 per cent mortality, even under very high-density conditions. Competition for space in territorial species is often of the contest type.

You may have noticed that while I started off using 'food' and 'resource' interchangeably, we've ended up talking about space, which is clearly not food. Barnacles may compete for space on a rocky shore, but they can't eat rock. This brings us to the question of what exactly is meant by 'resource'. Essentially, a resource is anything that a population can use up. This includes food, light, water and space, but not things like temperature and windspeed.

Thus describing a resource as being below its consumer in the food web is only correct for food resources; and even then the picture doesn't apply to plants, since food webs don't usually incorporate their 'food'.

Since I introduced competition as a biotic interaction, you may have become a little uneasy. How can an interaction with a piece of rock (taking a barnacle's-eye view) be described as biotic? Surely a piece of rock is about as abiotic as you can get? However, these worries largely disappear if we think of competition as an interaction among the competing individuals as well as between them and whatever resource is in short supply. With this way of looking at it, competition is a biotic interaction regardless of the nature of the resource.

Extinctions – Local and Global

This chapter started with a question: why is extinction so rare? We have now got to the point where we can readily provide an answer. An individual population is resistant to extinction because if its size begins to drop – thereby making it more vulnerable to extinction – the negative feedback effect of density-dependent processes tends to boost it back up to higher numbers. Equally, if numbers get *too* high, density-dependent forces tend to push them down again before total resource destruction and consequent extinction can occur. Density dependence, then, is the 'ecological thermostat' that makes *populations* resistant to extinction; but what of the species as a whole? We usually refer to the demise of a population as a local extinction, to distinguish it from the loss of an entire species (and so loss of all its constituent populations), which can be referred to as a global extinction. In fact, when we turn our attention to the latter, broader-scale phenomenon, it's clear that density dependence is again the main extinction-preventing agency.

Looked at in one way, this extrapolation is obvious. Since a species' total presence in the ecosphere is no more than the sum of all its local populations, its resistance to global extinction is clearly dependent on their resistance to local extinctions. However, the rarity of global extinction is aided by two additional factors: migration and asynchrony. I use the latter term to refer to the quasi-independence of ecological events happening in different parts of the world. If all of a species' populations exhibited marked troughs in population size at the same time, the probability of global extinction would be considerably higher than it is. Luckily, there's a high degree of asynchrony from place to place. If a population of a British insect species (say) is temporarily very low in numbers in one locality, it may well be that an adjacent – or at least not too distant – population is peaking. Of course, factors such as harsh winters can have a fairly widespread effect; but they do not override local processes to the extent that all populations vary in an identical temporal pattern, so there's always some degree of asynchrony.

Migration interacts with this asynchrony. At a given moment in time, any one species can be thought of as a patchwork of more-or-less permanent populations with a scatter of migrating individuals moving between them. Density-dependent processes cause the probability of successful establishment of migrants in an already inhabited area to be higher if the density of the residents there is low. So ultimately resistance to global extinction is based on the 'triple alliance' of migration, asynchrony and density dependence.

At present, the greatest threat to this rather efficient system is man. Because we destroy entire natural habitats (to replace them with agricultural ones or with housing or industrial estates), and because our activities are so widespread across the ecosphere, we have caused, and will continue to cause, both local and global extinctions. To many ecologists, the death of a species with a history measured in millions of years is a very tangible event: like a light that has shone for what seems like an eternity suddenly going out with no hope of ever being re-lit. Clearly, we should endeavour to reduce our 'speciecide' to a minimum. I'll return to this issue in Chapter 15. In the meantime, let's go on to look at a different kind of threat to a species – its closest ecological neighbours.

8

Doing Battle over a Limited Food Supply

If natural populations interacted only with their resources and the weather, an ecologist's work would be relatively simple. However, despite our concentration on those two things in the previous chapter for the sake of simplicity, we know that in reality a population's interactions are much more multifarious. Indeed, I've already equated ecological communities with interaction webs in which any species population may be said to interact – with varying degrees of directness – with most or even all of the other populations in the system.

We must, however, avoid the temptation to leap from an oversimplified approach to, in a sense, an overcomplex one. Many of the very indirect interactions, wherein the route from one species to another in the web meanders through a large number of intermediates, can to all intents and purposes be forgotten about with little adverse effect on our attempts to understand the overall stability of the community. For example, suppose that two species cohabiting a particular patch of woodland are an owl and a leaf scale insect. Further, suppose that the owl largely subsists on a diet of rodents which live at ground level and do not encounter the scale insect which is found in the canopy. Now it may be that the owl and the scale insect interact nevertheless, but in some very indirect way. For example, the rodents may consume pupae from the leaf-litter on the forest floor that belong to a species of insect whose adults, when they emerge from the pupae, ascend to the canopy and feed on the foliage of the trees, thereby potentially reducing the space available for colonization by scale insects. If the owl's population goes up, the rodent's may go down, the defoliating insect's up, and finally the scale insect's down.

Why can we be reasonably confident that a neglect of such indirect interactions will not be detrimental to our attempt to fathom the balance of nature? The answer is simply that beyond a certain degree of indirectness (which admittedly is difficult to pin down), the effect of one species on another is diluted out by so many other influences that it is negligible. Owl numbers *may* affect rodent numbers, but they are only one of several influences on the rodent population. The same compounding of multiple influences operates at each link *en route* towards the scale insect. The combined effect of all this compounding is that the owl population could probably double without causing a single scale insect to die. If so, then we would be wasting our time studying the 'interaction'.

Our approach, then, will be in yet another sense of the phrase, a 'middle way' – neither too comprehensive nor too simple. We will concentrate on the strongest and most direct interactions in which a species is involved (these two features usually varying in parallel) at the expense of the weaker and more indirect interactions. Of course, the cut-off is a subjective one, but we can at least modify it later if need be. That is, if the answers we get about community structure based on an analysis of reasonably direct interactions don't seem to make sense, we can alter our approach to be a little more inclusive.

What strong and direct interactions does the typical species population get involved in? The most obvious interactions, in most cases, are those involving feeding. Natural enemies (predators and parasites) and resources have a major impact on most populations. Of course, the nature of the interactions varies depending on your whereabouts in the food web. Top predators, by definition, have no predators, but they are most unlikely to have no parasites. Plants have no prey, but they do still have food resources in the form of light, water and soil nutrients. So while there is much variation in the *type* of natural enemy and the *type* of resource, all populations have both of these in some form or other, and in virtually all cases, interactions with these two classes of ecological factor are of prime importance.

As I mentioned earlier, most species are both consumers and resources, and consequently interactions with 'food' and 'enemies' are in a sense the same thing (trophic interactions) looked at from opposite points of view. We gave these trophic interactions some consideration in the previous chapter, and will examine them in more detail in the next. The present chapter is devoted to a kind of interaction that is not direct – as a predator/prey interaction is – but is only one level removed from it: a 'primary indirect' interaction, if you like.

What I'm referring to here is the interaction known as inter-specific competition, wherein two or more species (often but not always closely related) have a mutually detrimental effect on each other's populations through their joint use of a limited resource. Strictly speaking, this is exploitative interspecific competition. It's also possible for species to be mutually inhibitory in a more direct way such as through secretion of toxic substances that harm each other, in which case we have interspecific interference competition.

One way of looking at interaction webs – and therefore communities – is as compounds of horizontal and vertical interactions, the former being competitive ones, the latter trophic ones. Taking this view, an understanding of the balance of nature at the overall, community level should be reached by investigating the stability characteristics of the horizontal and vertical kinds of interaction separately, and then by looking at the relationship between them. This chapter and the following two are aimed at doing just that.

Of course, communities incorporate other types of interaction as well as competition and predation, and an obvious example of this 'other' category is mutualism (sometimes called symbiosis) in which both (or all) interacting species benefit from their association. Pollination of plants by insects falls into this category so it can hardly be said to be unimportant. However, because ecologists are primarily concerned with understanding questions of stability and balance, they've invested more effort in studying those interactions (competition, predation) which seem inherently un-stable, and hence would appear at first sight to threaten ecological balance, than they have in studying interactions such as mutualism where, in a sense, there is no stability problem. I will, here, echo this general bias.

Since we're about to focus on exploitative competition, it's worth noting – before we get embroiled in the details – that for such competition to occur in the first place some resource must be in short supply or, to put it another way, must be limiting to the populations concerned. For the moment, we'll simply assume that this is a common state of affairs. However, there is much current debate in ecology as to whether – at least in some groups of organisms – this is in fact true. We'll come back to this issue in the following chapter.

Making a Start

As always, we're concerned with looking for repeated ecological patterns and for general principles which may go some way towards explaining those patterns. One pattern of competition, which was pointed out by Charles Darwin in 1859,[13] is that the intensity of competition tends to become greater as the taxonomic relationship between the competing species gets closer. So we would, for example, expect congeneric species of birds to compete more severely, on average, than species from quite different families of birds. While this generalization is almost certainly true in a probabilistic sense, there are of course exceptions to it; and some interesting cases of competition between very distantly related kinds of organism are known – for example between seed-eating ants and rodents in desert systems, as described by the American ecologists Brown and Davidson.[14]

Because of this general relationship between taxonomic related-ness and intensity of competition, ecologists interested in studying competitive interactions have usually directed their attention to pairs or small groups of closely related, often congeneric, species. Some such studies have been based on populations living in the wild, others on laboratory microcosms, an example of the latter being Gause's study of competition in the genus *Paramecium*, which I introduced in Chapter 5.

Regardless of whether the setting is experimental or natural, the focus of interest has been – as in other parts of ecology – stability. The question has been posed: what enables some pairs of competing species to reach a state of stable coexistence while other pairs seem to undergo an intrinsically unstable interaction which results in one of them becoming extinct? The hope, of course, is that there is some *general* answer to this question, that applies regardless of whether we are talking about competition between birds, snails, insects, plants, *Paramecium* or anything else.

Let's look at a few case-studies of competitive systems before we think about whether any generalizations can be made about competitive stability. An appropriate starting point is Gause's work, since we've already had a glimpse of this; but later we must also consider competition between more complex kinds of organisms and in more complex (and more natural) environments.

Battles in the Laboratory

We had a preliminary look at interspecific competition in Chapter 5, and I mentioned then that the typical outcome of such competition in simple laboratory environments was instability. That is, one species comes to dominate, and drives the other to extinction over a period of several generations – a process known as competitive exclusion. In some case-studies, notably that by the American ecologist Thomas Park[15] on flour beetles of the genus *Tribolium*, all experiments gave this result. Which species won and which was excluded could be reversed in Park's experiments (by changing the temperature and humidity of the environment in which competition took place), but not a single experiment resulted in stable coexistence of the two species.

While the majority of laboratory studies of competition have yielded a result similar to Park's – that is, universal competitive exclusion – a few have resulted in stable coexistence under some conditions and exclusion under others. This is an informative outcome because it allows us to compare the nature of the environments and the species' usage of them in the two cases and so (hopefully) to work out what determines which outcome – coexistence or exclusion – occurs.

Gause's case-study with *Paramecium*, which we briefly encountered in Chapter 5, was one of these relatively rare cases with a mixed outcome. In his overall programme of work on competition in *Paramecium*, Gause used three species – *P. aurelia*, *P. bursaria*, and *P. caudatum*, (or a, b and c if you don't like these names). The experiments I described earlier were those involving *aurelia* and *caudatum*, both of which, Gause informs us, feed predominantly in the upper reaches of the bacterial suspension which he provided as their food resource. As we saw, the standard outcome of competition between these two species is exclusion (of *caudatum*). Interestingly, however, when either *aurelia* or *caudatum* competes with *bursaria*, which has a tendency to feed lower down in the bacterial suspension, stable coexistence ensues.

It must be emphasized that the idea here is not that *bursaria* has a non-overlapping depth distribution with the other two species and that there is therefore no competition between them. Rather, the idea is that they do overlap (see Figure 12) and competition still takes place, but coexistence is able to occur despite this competition because of the statistical difference in feeding depths.

It's convenient to have a phrase for differences of this general kind which can be used to describe differences in the feeding

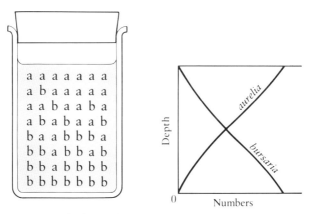

FIGURE 12 Presumed relative depth distributions of *Paramecium aurelia* and *P. bursaria* in Gause's competition experiments.

behaviour of closely related species of *any* sort of organism. The phrase we use is 'niche differences'. The British ecologist Charles Elton,[16], who was one of the first to promote this term, distinguished niche from habitat by saying that a habitat was an animal's address, a niche its profession. Since then, 'niche' has been used in various ways, but the predominant usage, which corresponds well to Elton's, equates niche with feeding behaviour. In the *Paramecium* system, then, *aurelia* and *caudatum* have a niche as 'top feeders', *bursaria*'s niche is as a 'bottom feeder'; but all are broad-niched, and consequently the niche of *bursaria* overlaps the others. However, since *bursaria* has a niche difference with the other two (statistical rather than absolute, as we already noted), it is capable of reaching a state of stable coexistence with either of them.

How do niche differences 'work' to promote stable coexistence? In order to answer this question, we must first recall the meaning of stability: the tendency of a system to return to an equilibrium value following a perturbation from it. Suppose for a moment that in a given tube of bacterial suspension, *aurelia* and *bursaria* settled down to relative frequencies of 60 per cent and 40 per cent, and persisted at those values for many generations. If you observed a result of that kind, you would probably suspect that those values represented an equilibrium. One way to test that suspicion would be to add or remove large numbers of one of the species, thus perturbing the relative frequencies. If, over a period of a few generations, they returned to their pre-perturbation starting point, then an equilibrium it is.

What causes the return to equilibrium following perturbation? Well, suppose that the perturbation takes the form of an addition of extra *bursaria* to the system. This causes some problems for *aurelia*, since it faces increased interspecific competition. But it causes a greater problem for *bursaria* itself, because intraspecific competition within that species is increased by more than its interspecific equivalent. You can think of this as extra *bursaria* inhibiting the population growth of its own species more than that of *aurelia* because, while it has some niche overlap with *aurelia*, it has total niche overlap with itself!

I think the way in which a niche difference functions as a

stabilizing mechanism is now fairly clear. If, in the face of perturbation – whether caused by an experimenter in a tube or a natural agency in the wild – one of a pair of competing species increases above its equilibrium, this will favour the other, so the system returns to equilibrium. If it 'overshoots' the equilibrium, then the advantage passes to the other species, and the direction of change in species proportions is reversed. So the system can be thought of as continually 'seeking' its equilibrium. The actual value the equilibrium takes is of no particular consequence. If a tube with parallel sides results in an equilibrium with 40 per cent *bursaria*, a conical flask would presumably produce an equilibrium value for *bursaria* in excess of this figure, with the magnitude of the excess depending on the degree of 'conicality'.

To strengthen our claim that niche differences promote competitive stability in general, it would be nice to have an example where the system works in a similar way to Gause's *Paramecium* system but involves completely different kinds of organism competing for a completely different resource. A programme of experiments that I conducted in the early 1980s, with two species of *Drosophila* fruit flies, satisfies both these requirements.[17]

The species concerned were the familiar genetic work-horse *D. melanogaster* and its larger and more darkly-pigmented cousin, *D. hydei*. The resource I provided for these two was a standard *Drosophila* medium contained in glass jars, which were screwed into the undersides of perspex boxes to form 'population cages'. The larvae live within the food medium, while the adults fly around in the perspex box and, whenever food bottles are replaced, descend to the surface of the fresh food and lay eggs on it. In most other systems of this kind, one species of *Drosophila* competitively excludes the other – the typical laboratory result. However, with this particular species pair and this particular environment, stable coexistence occurs, provided that a reasonable depth of food is present. The mechanism underlying the stability is a niche difference between *melanogaster* and *hydei* which, coincidentally, takes a rather similar form to that between Gause's *Paramecium* species in that it involves depth of feeding within the medium.

One gap in Gause's work is that he didn't provide us with any

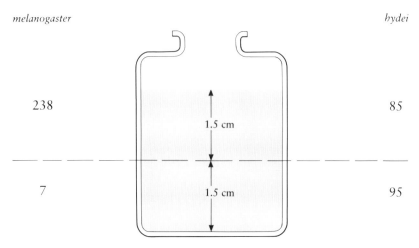

FIGURE 13   Relative depth distributions of larvae of two species of *Drosophila* in a resource bottle.

data to back up his description of the various species as top or bottom feeders. Since I was keen not to leave a similar gap in my story about competition between *melanogaster* and *hydei*, I sectioned the food medium into layers and looked at the number of larvae of each species in each layer. The result: lots of *melanogaster* at the top and very few at the bottom, with *hydei* showing an opposite (but much less marked) preference for the lower reaches (see Figure 13).

If a niche difference in larval feeding depths is indeed the stabilizing mechanism leading to coexistence, it should be possible to de-stabilize these systems by providing a shallower resource unit. To test this, I ran another series of experiments with an insufficient depth of food to permit the species to separate out. As predicted, the result was competitive exclusion (of *hydei* by *melanogaster*) in all replicate cultures.

Battles in the Wild

Up to now, we have a fairly neat and consistent picture of competitive stability. If the competing species have similar niches, one competitively excludes the other, and which one wins depends on the prevailing environmental conditions. Given a niche difference, however, a state of stable coexistence is reached. Just how much of a niche difference is necessary to have a stabilizing effect we'll examine later. But now I want to turn our attention outward from the laboratory to nature. Does the neat and consistent picture developed from laboratory-based case-studies apply also to their natural counterparts?

In fact, a curious state of affairs prevails in our attempts to understand competition in nature. For a start, we cannot often observe particular competitive systems which in some cases give rise to exclusion and in others to coexistence – so attention usually focuses *either* on an apparent case of instability and exclusion (such as the red and grey squirrels in Britain) *or* on an apparent case of stability and coexistence. Since most studies have been of the latter kind, I'll concentrate on these.

In this kind of study, the primary observation is usually that two or more closely related species coexist in a particular place at a particular time. The three key questions of interest, in their logical order, are as follows: (a) are they competing? (b) is their coexistence stable? and (c) what kind of niche difference or other stabilizing mechanism is operating? If the answer is 'yes' for either of the first two questions we proceed to the next one; if it is 'no' then the next question is inappropriate.

I would venture two guesses about competition in nature. First, I suspect that 'triple positives' abound. That is, I suspect that there are very many cases of stably coexisting competitors where the stability is generated by a potentially identifiable niche difference. Second, as far as I am aware there is not a single case study in which we can conclusively answer all three questions. Let's briefly examine two case-studies – one well known, the other not – which in different ways illustrate the problems encountered by ecologists interested in understanding competition in nature.

The first, lesser-known, study involves seaweed flies of the genus *Coelopa*. Two species are present around the coast of Britain – *C. pilipes* and *C. frigida*. Their habitat is 'wrack beds' – expanses of rotting seaweed deposited on beaches. In some ways this is a simple habitat; it's a sort of brown goo and, past a certain degree of decomposition, has a superficial resemblance to some *Drosophila* media used in the laboratory. The two species are found together in many (perhaps most) of these wrack beds, which constitutes our primary observation, as already noted. We then turn to the three 'key questions'.

The enormous densities of larvae found in many wrack beds make it almost impossible to believe that the species are not in competition; and the case for competition is strengthened by the detection of a 'stressing' effect of *pilipes* on the genetic system of *frigida*[18] (though evidence for a reciprocal effect is lacking). The coexistence is probably stable, since there are particular areas in which coexistence has been noted over periods of more than a decade – though we have to admit that such an observation is also consistent with a very slow trend towards competitive exclusion.

Turning to the question of what niche difference underlies the stability, we come up against the bewildering variety of nature. In the laboratory, the simplicity of the systems concerned doesn't permit much other than a difference in feeding depth. But consider the possibilities in the natural *Coelopa* system. The species could differ in how decomposed they like their food, in which species of seaweed they prefer, in the degree of aeration or humidity of their feeding locality, or even in depth of feeding like the laboratory systems we looked at earlier. And this list is far from exhaustive.

At the present stage in our understanding of the *Coelopa* system, we haven't the faintest idea which of these (or lots of other) niche differences prevail. So to come back to the three key questions, we are pretty sure that competition is going on, we think that the coexistence is stable, but we are completely ignorant of the stabilizing mechanism.

Let's now turn to the other, better-known, case-study – that by the American ecologist Robert MacArthur[19] on warblers of the genus *Dendroica*. In coniferous-forested areas of New England, up to five of these species can be found living in the same place – again, the primary observation. Since all are insectivorous, they are potential competitors. MacArthur showed, however, that there were differences (again statistical rather than absolute) in the foraging sites of the different warblers. Each concentrated on different zones within each tree, as shown in Figure 14.

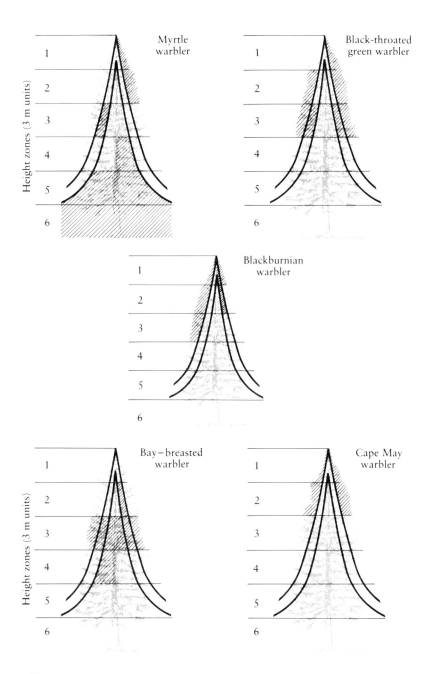

FIGURE 14   Feeding zones (cross-hatched) of five species of warblers in coniferous forest. The patterns on the left are based on % of total observation time; those on the right are based on % of number of observations.

Such a niche difference could indeed act as a competitive stabilizing mechanism providing, of course, that the insects caught in the different foraging zones were not all freely mobile between them. However, it's again sensible to take our three key questions in their logical order. With regard to the occurrence of competition, we must admit to total ignorance in this case. Whether resources are ever actually limiting, and whether if one warbler species was experimentally removed any of the others would increase in density is unknown. The case for stable coexistence is, as in *Coelopa*, based on co-occurrence being observed over an extended period, which is not conclusive. And on niche differences – well, we know of one, but whether it is the only or most important one is an open question.

Notice that neither in *Coelopa* nor *Dendroica* studies is there a single certainty. We think or surmise that competition occurs, we suspect that the coexistence is stable, and we either have no concrete clues about niche differences (*Coelopa*) or we can identify one (*Dendroica*) about whose importance we are uncertain. This rather unsatisfactory state of affairs is an adverse comment neither on the choice of case-studies nor on the competence of those conducting them; but rather is a comment on the enormous difficulty of demonstrating anything with certainty in the multivariate messiness of nature.

The Rules of Battle

It's time to remember where we're going. At the end of the opening section of this chapter, I said that before looking at the general principles of competition, we should examine some of the case-studies of individual systems from which those principles have arisen. Having now done this, it's time to turn to the principles. What we need to do is to abstract out of the various case-studies the essence of what is going on – especially in relation to stability – and present it in some way that is independent of 'local peculiarities', that is, species-specific or environment-specific details. I've already made a start in this direction by introducing the niche concept; and I'll now attempt to build on this concept so that we end up with, essentially, an abstract theory of competition.

The first step is to picture niches in some abstract graphical manner. This is often done by drawing what we call resource utilization curves, as shown in Figure 15 for two competing species. Here, the resource is assumed to be limiting, and the usages of it by the species concerned are plotted against some variable by which the resource may be described. It may be depth within resource units, degree of decomposition, size of units (e.g. seed size for granivorous birds), or whatever. Some degree of separation of the curves (as shown) can be described as a niche difference, and may function to promote competitive stability in the manner already described, namely each population inhibiting its own growth more than that of its competitor. Another way of describing this state of affairs is frequency dependence, since it involves a decrease in the fitness of each species as its own relative frequency in the two-species mix increases. There's an obvious analogy here with density dependence in single-species systems.

One assumption that is hidden in the apparently symmetric picture of competition shown in Figure 15 is that items of food at various points along the 'resource variable' are independent of each other. This would be true, for example, if species A and B are granivorous birds and the resource variable is seed size. However, it would not be true if A and B were the two *Coelopa* species and the resource variable was 'degree of decomposition of seaweed',

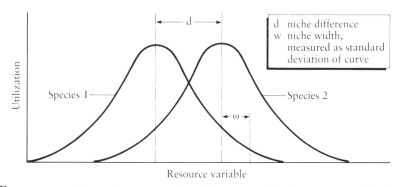

FIGURE 15   Schematic representation of resource utilization curves (niches) for two competing species.

because removal of resource at any point on the x-axis would then have a depleting effect on all points to its right. Here, dependence and asymmetry are obvious. There are other cases where the situation is intermediate. For example, in *Drosophila* food bottles, removal of 'top' food does not cause food at the bottom to disappear, but it reduces the depth available, and consequently alters the status of the food that is left. One of the most gaping holes in current competition theory is that the effects of varying degrees of non-independence of resource have not been adequately investigated; as evidenced by the fact that some ecologists can still be found proposing a 'temporal niche difference' as the cause of competitive stability, when it will often instead act in a destabilizing way by giving a straight competitive advantage to the earlier-feeding species.

In those cases where there are no such problems of asymmetry, it is thought that a certain degree of niche difference will lead to stability. This proposal is often known as the principle of limiting similarity. However, identifying what the critical degree of difference is is a difficult business. Those ecologists who have pursued this issue have used mathematical models to some avail. One key result of this work (which derives from a seminal paper by ecologists Robert May and Robert MacArthur in 1972[20]) is that there is no universally applicable degree of niche difference that will produce stability in all cases. Rather, the necessary degree of separation between the species' niche maxima is dependent on the unpredictability of the system: greater unpredictability requiring greater separation if stable coexistence is to be maintained. Although the result 'falls out of the mathematics' it corresponds with most ecologists' intuitive feeling that tighter packing of species should be possible when conditions are more predictable. It also, as the British evolutionist John Maynard Smith has pointed out, gives us "a first clue to understanding species diversity".[21] But this is jumping the gun. Before we ascend to the community level, we need to look at the other major type of ecological interaction – predation.

9

To Eat and To Be Eaten

All organisms eat. Their food may be flesh, blood, leaves, roots or sunlight, but if they don't eat *something* they won't be around for long. Equally, all organisms are eaten. They may be consumed whole, dismembered, nibbled at or parasitized, but it is unlikely that any escape. The natural world can be thought of as a never-ending struggle to acquire as much energy as possible and to avoid giving it up to anyone else except one's offspring. From the viewpoint of a natural population, then, its interactions with its resources and its consumers are of the utmost importance. While I've acknowledged this before, I have, nevertheless, concentrated very much on the 'eating' rather than the 'being eaten' side of things. In the previous two chapters we first envisaged a population being regulated primarily by density dependence arising from intraspecific competition for food; and we then extended this notion of competition to two or more species whose diets overlapped to some degree. We ended with the rather elegant idea of 'limiting similarity' which may provide an important insight into community structure.

All this, however, is conditional on populations being regulated by their resources. That is, the reason that per capita growth rates become negative above some (equilibrium) density is that shortage of a resource decreases survival and reproduction probabilities to below maintenance levels. Yet it's possible to envisage a situation wherein, although population size does indeed equilibrate, resources are actually 'superabundant', meaning that there is vastly more than enough to feed the population concerned. What we need to ask at this stage is "what sort of observation would suggest that this is happening?" and "how could such a situation arise?" I'll deal with these two questions in turn.

Suppose you are monitoring the population size of some small invertebrate that is rather specific, in its feeding habits, to a particular food plant. Further, suppose that you have monitored this population over many generations and have found a very clear pattern of constrained fluctuation, similar to the patterns we saw for the (fictitious) thumper and the (real) flycatcher in Chapter 7. Now if, throughout the study period, the food plant population had been widespread and dense, and had consisted of vigorous individual plants with luxuriant foliage, you'd be forced to think very carefully about whether the above state of affairs – an equilibrial but not resource-limited herbivore population – might prevail in this particular case. Certainly, the pattern of constrained fluctuation strongly suggests the existence of an equilibrium. But equally, the continual availability of excess resource renders competition for food an unlikely source of the equilibrating force – i.e. the density dependence. Of course, you'd have to be a little careful about drawing a firm conclusion. After all, the herbivore might be dependent on a particular sort of plant tissue which, despite the luxuriant growth, was in short supply. But if this were found not to be the case, then you'd have little choice but to look outside the realm of resources to discover the cause of the equilibrium.

In fact, there are many examples of this 'just suppose' scenario, and I'll single out one of them for discussion. This is the case of bracken-feeding insects, studied by the British ecologist John Lawton.[22] Surprisingly, perhaps, the bracken that is a familiar sight to the British hill-walker is found throughout the world as a single species (*Pteridium aquilinum*). It is therefore, like man, a species that can be described as cosmopolitan. Yet throughout its range it comes into contact with a bracken-feeding insect fauna which differs quite considerably from place to place.

Lawton made a number of interesting observations on this system. First, the bracken rarely if ever is decimated by its insect consumers. Second, despite there being much variation in the number of consumer species from place to place (from three in

Arizona to fourteen in Papua New Guinea), this variation seemed to have little if any effect on the densities of the species that actually occur at any one site. Finally, the niches of the co-occurring consumers are often very 'clumped', rather than being regularly dispersed as the theory of limiting similarity would predict.

On a hypothesis of exploitative competition, none of these observations make sense. However, if we're prepared to acknowledge the possibility that populations may be regulated around an equilibrium density by a factor other than their resources, then perhaps things will begin to fall into place. This brings us to the second of the two questions raised earlier, namely "how could such a situation arise?"

Actually we've already had a clue to the answer. I mentioned in Chapter 7 that the main sources of density dependence were biotic interactions, and we've seen that the most direct biotic interactions that a species undergoes are those with its resources and its consumers. Might it not be, then, that some populations are regulated around a rather low equilibrium by the action of one or more of their 'natural enemies' – i.e. predators and parasites? The answer is 'yes', and exactly how this happens is the problem that we now need to address.

Predators and Density Dependence

As usual, it helps to picture something fairly specific, so let's not include all 'natural enemies' in our discussion. Instead, I'd like to focus on predators in the narrower sense of that term, thus excluding herbivores and parasites. So we're talking about animals eating animals. Yet this is still a broad enough category to include the bracken system (where the prey and many of the predators are insects), as well as lions eating zebras, birds eating snails (of which more shortly) and all other cases of carnivory as well.

We need, at this stage, to appreciate both the similarities of, and the differences between, predator-determined and resource-determined equilibria. To facilitate this, let's concentrate our attention on a species of snail that inhabits a variety of environments. I have in mind the banded snail *Cepaea nemoralis*, which is found in sand dunes, woodlands, grassland areas, hedgerows and even, occasionally, in gardens. Now suppose that we focus our attention on two populations living among rather similar ground flora (a mixture of grasses and broad-leafed species), but one in a semi-wooded locality, the other in an entirely treeless one. We'll imagine that avian predators such as thrushes are present in the semi-wooded locality (because of the presence of suitable nesting sites) and absent from the other.

If we could repeatedly estimate the per capita net growth rate (PCNGR for short) in both snail populations over a longish period of time during which densities fluctuated considerably, we might obtain a result like that shown in Figure 16. The right-hand scatter of points, representing the open habitat, can be interpreted as fluctuation around a resource-determined equilibrium (E_R) with intraspecific competition providing the source of the density dependence that brings the equilibrium into being. The left-hand scatter of points represents the wooded habitat – the one with avian predators – and can be interpreted as fluctuation around a predator-determined equilibrium (E_P).

This is not to say that the population in the open habitat suffers no predation. Rabbits may well be present, and are known to

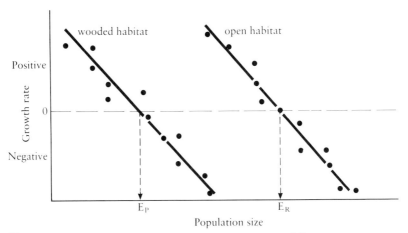

FIGURE 16 Different equilibrium population sizes in different environments. (E_P predator-determined equilibrium; E_R resource-determined equilibrium.)

consume *Cepaea*, though not in vast numbers. Rather, what we are saying is that predation on that particular *Cepaea* population is weak and intermittent, and exhibits no detectable relationship with density. Consequently, the population of snails keeps increasing until competition for resources begins to bite, at which point it begins to stabilize. If we were able to identify the precise limiting resource (which is often impossible), we should find that it's very thin on the ground. An equilibrium is reached, as we've seen before, when the shortage of this resource is such that it causes the PCNGR to become zero.

How does the PCNGR drop to zero long before that stage in the more heavily predated population? To put it another way, how can a predator act as a density-dependent factor? Our picture of how competition for resources acts in this way revolves around the resource being used up; but you can't 'use up' a predator, and even if you could, this would be good news, not bad news, so surely there would if anything be a positive feedback system, causing runaway population growth rather than stabilization?

In fact, there are at least three ways in which predators can cause density dependence of PCNGR in their prey, and we'll look at these in turn. For simplicity, I'll pretend that predators only cause mortality – so we can talk specifically about density-dependent mortality rates rather than PCNGRs. However, it should be appreciated that predators can influence prey natality as well.

The first source of density dependence is what's referred to as the predator population's 'numerical response' to increasing prey density – namely the ability of predator numbers to increase if prey numbers increase so that the predator 'tracks' the prey. Whether or not this causes density dependence of prey mortality depends on just how rapidly the predator population responds. On its own, this mechanism is unlikely to be effective in many cases. It will only 'work' (i.e. cause density dependence) if the predator population expands more rapidly than that of the prey, which seems unlikely. So we turn to other factors which, when operating in conjunction with the numerical response, provide a more convincing mechanism for density dependence.

One of these is the prey 'refuge'. We first came across this concept in Chapter 5 in relation to Gause's laboratory microcosm of predator-prey systems, where *Paramecium* was consumed by *Didinium*. In that case, the refuge altered the outcome of the predator–prey interaction, but did not cause stability. However, in other situations refuges may well have a stabilizing effect. In the *Cepaea*/thrush system, for example, it's easy to imagine this happening, and the process is illustrated in Figure 17. The rationale goes something like this. Snails feed on the topsides or undersides of leaves. Those on top are more easily visible and have a much greater chance of being eaten. The snails feed in the most protected microsite available, but as their density increases, a greater and greater proportion of the population has to forage in suboptimal sites. Therefore, as density increases, so does the percentage mortality caused by predators.

So far, we've considered only a single predated population – the one in the semi-wooded locality. We have treated it 'in isolation' – pretended, if you like, that both the snails and the thrushes occupy a discrete spatial area and have no contact whatever with neighbouring populations. Yet in reality, such a clear-cut delimitation is unusual. What's more commonly encountered is an area where snail density is high surrounded by areas where it is low, beyond

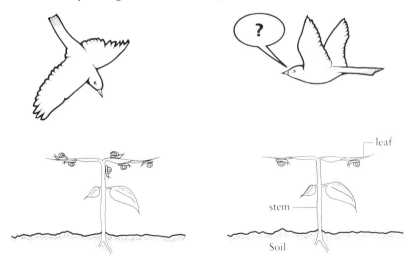

FIGURE 17   Refuges from predators are not so easy to find in high-density conditions: illustration for a bird predator/snail prey situation.

which density increases again; and indeed this sort of pattern is often found on more than one spatial scale.

Confronted with a patchy distribution of this kind, what will a predator do? Well, it isn't hard to answer this question, especially if you have some experience of being a predator yourself – for example as an angler. As you get to know a lake or a stretch of river you realize that fish density is usually higher in some places than others, so you head straight for the 'likely spot'. While we cannot get inside the brain of a thrush, it seems unlikely that they would behave differently. So if prey density increases in one 'patch', or local population, it will suffer heavy predation, while areas that are currently sparsely populated by prey will be left alone. Thus each local population of prey will undergo density-dependent predator-induced mortality, and may, as a consequence, fluctuate around an equilibrium level below that which its food supply could sustain.

So far so good. We've got as far as thinking our way through the problem of how predation might act to cause density dependence and hence to stabilize prey populations. But is there any evidence that predators actually do behave in this way? After all, a predator-mediated equilibrium is at this stage only a possible explanation of John Lawton's findings with bracken-feeding insects. And while thrushes do indeed eat *Cepaea* (after smashing their shells on a convenient stone or 'anvil'), our discussion of that system was very much in the mould of a thought experiment rather than an analysis of any real data.

Actually, we *do* have plenty of data-sets revealing the density-dependent action of natural enemies. One example comes from a well-known case study on the winter moth (*Operophtera brumata*). This species gets its name from the fact that the adults are, unusually, active in winter, while the larval stage occurs in spring, and the insects 'rest' as pupae from about June to November, before the next generation of adults appears in December. Like all insects, the winter moth suffers attack from a variety of predators consuming various stages of its life-cycle. When predation-induced mortality of pupae is plotted against density of pupae (or density of the larval stage immediately preceding pupation), a clear positive association – i.e. density-dependent mortality – is

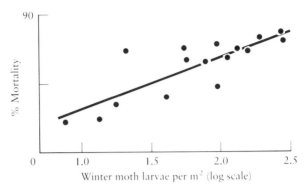

FIGURE 18   Density-dependent pupal mortality caused by natural enemies in the winter moth, *Operophtera brumata*.

found,[23] as shown in Figure 18. So not only can we envisage mechanisms through which predators could act in a density-dependent fashion, but we can also observe them to do so. This opens up the possibility that a predator operating in this way could cause PCNGR to go negative long before a deficiency in the food supply would cause this to happen through intraspecific competition. But which situation is commoner in nature?

Two Views of Nature

In this section, we'll ultimately get to what I consider to be one of the most important current debates in 'pure' ecology. But the issue is complex, so we'll get there in stages. Our first step is to ask the question: can any ecological factor other than resources and natural enemies act as a source of density dependence? The answer to this question is a qualified 'no'. (The qualification will have to wait until Chapter 13, I'm afraid.) If I'd used the phrase 'food and predators' instead of 'resources and natural enemies', then of course the answer would have been 'yes'. Anything a population needs and that it can potentially use up (including space), and any attacking agent that can respond to a population's increase with a disproportionate increase in the severity of its attack (which includes parasites), can be a source of density dependence. So we have to use the broader categories of resource and enemy to include all of these. But having chosen the broader categories, we have now included *all* the sources of density dependence which a population may experience.

Naturally, there will be objectors to such a sweeping statement. In particular, it might be argued that climatic factors could, under certain circumstances, be density-dependent. For example, a small invertebrate might only survive in microsites with a very high humidity. The commonness of these sites will depend both on the overall climate in the region and the structure of the habitat – more dense foliage generally providing more humid conditions. It is possible to envisage a population of insects (say) that live among the foliage being limited in a density-dependent way by the availability of microsites with a suitable humidity. However, while climate is certainly involved here, it acts via the resource of space. And in general it may be said that, while various factors can influence the availability of resource or the abundance of natural enemies, it is still ultimately only these two groups of factors that can cause density dependence to occur.

We now turn to the question: what is the relationship between density dependence and population regulation? At one level, the answer is simple, and indeed we have already discussed it in Chapter 7: in a closed system, density-dependent natality and/or mortality is necessary if there is to be regulation around an equilibrium population size. However, some caution is necessary here, because if we find that a particular factor is indeed density-dependent (such as pupal predation in the winter moth example), we cannot immediately claim that this is the population's 'regulatory factor'. The problem is, of course, that more than one factor may be density-dependent in any given population. For example, it may be that three natural enemies, acting on the same or different life stages of an insect population, are all density-dependent over the range of densities through which that population normally fluctuates. It is then inaccurate to talk of a single regulating agent. Rather, what regulates the prey population is the downward slope of PCNGR values, which at a certain point on the scale of ascending density passes through PCNGR = 0. This slope is contributed to by three separate factors (in our example), none of which, in isolation from the others, is 'regulating' the prey population – rather it is their conglomerate action that regulates.

The next question is: what is the relationship between density dependence stemming from resources and that stemming from natural enemies? We have just seen that several entities in the enemy category can jointly regulate. Is it also possible to have joint regulation across categories – e.g. involving both a food resource and a predator? I think that the answer here is 'yes, it's possible, but it's rather unlikely', and I'll spend some time explaining this view.

Let's perform an imaginary field experiment. We'll return to the snail/thrush system, although the experiment would in practice be rather difficult to perform on that system. Suppose we take the population of snails living in the non-wooded habitat, which is fluctuating around a resource-determined equilibrium. In our earlier look at this population (Figure 16), we viewed the pattern of density dependence as being linear. But in reality a more complex pattern is likely to be found. One shape the relationship may take is a plateau followed by a convex curve, as seen in Figure 19. The idea here is that snail density can ascend a long way up from zero before resource limitation begins to be felt, but that

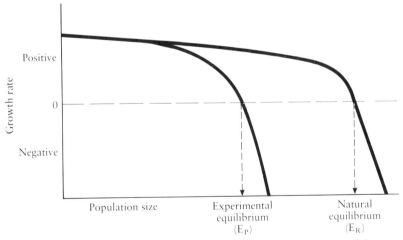

FIGURE 19   A possible effect on equilibrium population size of the experimental introduction of a predator.

once a resource becomes scarce, this brake on population growth bites increasingly hard if density continues to increase.

Now suppose we introduce into this local area a thrush population, having provided some sort of nesting facility to enable it to persist on a long-term basis. One possible outcome of this experiment is shown in Figure 19: the density-dependent mortality associated with the predator causes the population to reach a lower equilibrium (E_P), as happened in the wooded habitat in our earlier discussion. Assuming for simplicity that no other ecological factors than thrush predation and limitation by a particular resource (food, say) can conceivably behave in a density-dependent way in this particular population, it's clear that the predator is *the* regulating agent. Food can be excluded, because over the range of densities through which the population now fluctuates it is not a significant density-dependent factor – i.e. we are in the plateau region of the right-hand curve.

This experiment, then, indicates non-overlap of the ranges of density within which food and predator exert a density-dependent brake on population growth. If this situation prevails, then populations do indeed have a single regulating factor in the sense of it being either food or predator – and which it is depends simply on whether the predator is there or not. However, were the left-hand curve in Figure 19 to have been much closer to the right-hand curve, then food and predators could have jointly regulated the population. Undoubtedly this must happen in some cases, but I suspect that it's relatively rare: given the immense range of *possible* densities, the probability of the two types of factor (resource and enemy) causing a strong density-related drop in PCNGR over the same band of values seems fairly low.

We now have all the ingredients for a major dispute. There are only two sources of density dependence – resources and enemies. Since density dependence is required for regulation, there are therefore only two 'pure' kinds of regulation in natural populations – resource-mediated and enemy-mediated. While a mixed mode of regulation is possible, it may be relatively rare, as the two types of factor will tend to bite over different density ranges. So the question naturally arises: does 'upward' or 'downward' regulation prevail in nature? I use these labels because they are less cumbersome than 'resource-mediated' and 'enemy-mediated'; the directions up and down refer, of course, to the food-chain context.

The view that upward regulation predominates was popular until the late 1970s – and continues to be so in some quarters. Ecologists who have held that view include David Lack, G. E. Hutchinson and Robert MacArthur. Over the last decade or so, however, many ecologists have rebelled against this perception of nature, and have put a contrary view. For example, the American ecologist Stuart Pimm has stated that 'donor control' is rare in nature.[24] (Pimm uses donor control to mean resource-mediated or upward regulation, with 'donor' meaning energy donor and hence *food* resource.) John Lawton, referring to what I here call downward regulation and consequent lack of competition, argues that "a clear majority of the world's biota must behave in this way."[25] Others have urged compromise. A paper published (before its time) by three American ecologists[26] suggested that downward regulation may prevail in some trophic levels (notably herbivores), upward in others. The two purist views and this particular compromise are illustrated in Figure 20.

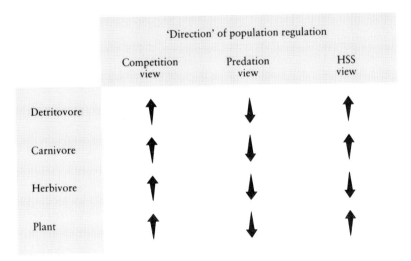

FIGURE 20 Three views of the prevailing direction of population regulation: the two 'purist' views, and the compromise urged by ecologists Hairston, Smith and Slobodkin.

Sequences versus Loops

One of the problems in thinking about ecological systems is that everything is always busy interacting with everything else. Even restricting our attention, as we are now, to an isolated food chain (which as we saw earlier is something of an over-simplification), most populations in the chain are both consumers and resources. This raises an interesting question which has remained hidden up to now but which is brought out of hiding by the compromise view of the flow of regulatory influences in food chains shown in Figure 20. The question is this: what implication does the direction of regulation in a given population have (if any) for its immediate upstream and downstream neighbours in the chain? For example, if a population in trophic level 3 is resource-regulated, does that mean that its resource (level 2) can be, must be, or must not be enemy-regulated?

In most of the foregoing discussion, this sort of question was lurking in the background. For example, what regulates the bracken populations in John Lawton's study? Or what regulates the thrush populations in our snail/thrush system? These questions now emerge because in Figure 20 we see both *sequences* – one-way systems, if you like – and *loops* of regulatory influence, and the contrast leads us to enquire whether both are indeed at least theoretically possible, quite apart from what actually goes on in nature.

This is an area in which progress can be made by building simple mathematical 'models' of interacting populations. You don't have to be a mathematician to understand the modelling approach, especially now that most of the mathematical steps are performed by computers. An ecologist interested in modelling predator–prey systems might proceed as follows. First, he devises an equation for the prey population. This takes the form: number of prey animals in generation X = number of prey animals in generation $X-1$ plus mathematical expression 'N' minus mathematical expression 'M', where N represents natality and M mortality, some or all of which is due to the predator. Next, he devises a comparable equation for the predator population. Then he supplies starting numbers for each population in generation zero (say 100 prey and 10 predators), feeds the whole thing into a computer, and asks it to draw a graph of population sizes over time for the next 100 generations. From this, he can see what equilibrium level (if any) each population settles down to.

The answers that we get from these models depend on how the expressions N and M are constructed, which differs from one model to another. Some models will allow both predator and prey to stabilize only when the prey is itself given a resource-mediated equilibrium. This, then, is a *sequence* (an upward one) of regulatory effects. Other models permit the predator–prey system to 'self-stabilize' without any extrinsic limitations such as resource-limitation of the prey population. This is a *closed loop* system. Since the models show that both arrangements are theoretically possible, we end up focusing, once again, on what arrangement actually prevails.

We're rapidly getting into the realms of guesswork now, since not only is there no consensus on what generalization we can make about nature, but there isn't even a consensus on the level to which that generalization should apply. Pure science, in any field, can be seen as a striving to achieve the most global generalization that is consistent with reality. So in relation to the direction of regulatory influences on populations in nature, we should like to be able to make a statement that all (or, more realistically, the vast majority of) regulatory influences work in an upward (or downward) manner. If this turns out to be untenable, which I believe it will, then we move down one level of generality from communities to trophic levels, and assert that within any given level all (or most) regulatory influences flow in a particular direction. The compromise view in Figure 20 is an attempt to do just that. If, even at this lower level, our claimed generalization is incompatible with reality, as it probably is, then we move down one more level, bringing us to 'guilds' – small groups of species utilizing given groups of resources, such as seed-eating birds or bracken-feeding insects. My own suspicion is that only at this level will it be possible to generalize about the direction of population regulation. However, the negative flavour of this relatively downbeat prediction is offset by the fact that density-dependent regulation from *some* source is a near-universal truth in ecology, and indeed may be the single most important strand in the balance of nature.

In order to make a start in understanding a complex subject, I have, in this chapter, reverted to the simplified concepts of food chains and trophic levels. Yet we've already seen in earlier chapters that these are at best great over-simplifications of natural systems, where an irregular and complex web of interactions is the norm. How are our conclusions from the present chapter (which implicitly assumes monophagous predators) affected by an acknowledgement that in reality most predators are polyphagous, and most prey species 'multi-predated'? The short answer, at present, is that we simply don't know. Yet observations and experiments on the population dynamics of multi-species communities characterized by food webs – indeed interaction webs – are now gaining ground. It's thus possible to begin to approach our key issues of the balance of nature not just in laboratory microcosms and in individual wild populations mentally dissected out from the rest of the system, but also in the undiluted complexity of a local chunk of ecosphere. It's to such community-level studies that we now turn.

10

RECONSTRUCTING THE WHOLE

Whole communities are notoriously difficult to deal with. They provide, in a sense, the ecologist's goal, since they are what we ultimately seek to understand, and indeed to conserve. But discussion of community-level issues in ecology carries with it a great danger of becoming confused by the plethora of interactions that we find, and by the almost limitless choice of possible approaches to the study of communities that is open to us. Bearing this danger in mind, I intend to be ruthlessly selective. That is, I'll channel the discussion in very specific directions and take an in-depth look at selected questions about communities, but will omit others entirely. So we shouldn't end up being vague and confused. We may, of course, end up instead having hacked a very clear path to what turned out to be the wrong destination, but I'd rather run that risk than the risk of becoming generally muddled.

A sensible starting point, given the complexity of the subject, is to define our terms. Of course, by this stage we have an intuitive feel for what a community is; but it may nevertheless be wise to invest a little effort in making such intuitive feelings more precise. It's also important that a community is not defined in a way which leads to futile discussions of whether communities 'really exist'. Clearly, communities are neither completely random collections of only very loosely interacting species, nor totally cohesive, ultra-homeostatic 'superorganisms'. Rather, they fall at some point on the continuum between these extremes, and I don't want to echo previous ecological debates about the precise location of this 'point'. (Indeed 'range' would be a better term, since some communities are more cohesive than others). So we'll simply define a community as 'the total biota of a given region' and have done with it. This definition implies nothing about how cohesive communities are, and leaves us in no doubt about their reality.

Three important distinctions now need to be made. First, that between community and ecosystem. The latter includes abiotic substances such as air, water and rock, while the former does not. We'll be concerned, as usual, with the biotic side of things. Second, we should distinguish between 'natural' communities and 'the rest' – including agricultural and other human-influenced systems. We'll concentrate here on understanding nature in its 'pure' undisturbed form, leaving the man–nature interface for subsequent chapters. Third, we should not confuse communities with community *types* (the latter also being known as biomes). Tropical rainforest, coral reef and arctic tundra are examples of community types or biomes, a broad classification of which is given in Figure 21. On the other hand 'a community' is a particular local patch of one biome. Some biomes naturally come in patches (lakes, ponds, mountain-top alpine systems), while others (forest, grassland, marine) tend to be more continuous. However, even in the latter case, we may still arbitrarily delineate a community unit for study, and in some cases the delineation may not be so arbitrary – e.g. a portion of coniferous forest almost enclosed by the meandering tributaries of a river system, which will form a barrier to many of the community's inhabitants. There is in any case nothing in our very broad definition of a community that requires it to be 'closed'.

It's generally agreed that most parts of the ecosphere, if left alone by man and other agents of disturbance (e.g. fires), would be characterized by a particular community type. If all farming ceased in Cheshire, for example, we wouldn't expect to end up, a couple of hundred years later, with tropical rainforest or tundra. (In fact, mixed deciduous forest would prevail.) Equally, if an area of coniferous forest in northern British Columbia was destroyed by fire, we wouldn't expect to end up with savannah. (Ultimately, coniferous forest would reappear.) These typical end-point community types characteristic of particular areas are sometimes referred to as climax communities, to distinguish them from the

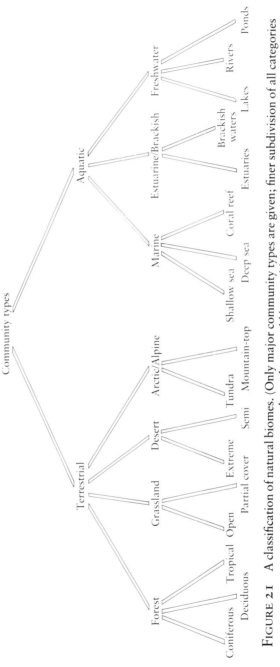

FIGURE 21 A classification of natural biomes. (Only major community types are given; finer subdivision of all categories shown here would be possible.)

series of ever-changing community types from which they arise, which are called successional communities. In this chapter, I'll concentrate for the most part on the climax state, but will consider succession in the next.

Community Stability

As always, our central concern is understanding the balance of nature; and now that we're dealing with nature in its undiluted form, we should, in theory, be able to address the issue of its balance most directly. However, care is needed because while at the population level there is an obvious choice of variable – the population size – whose stability or balance may be investigated, at the community level there are more alternatives from which to choose. And we have a duty to make this choice. Any approach in which we talk in general terms of 'a community' as being stable or otherwise is likely to lead into just the sort of muddle that I'm trying to avoid. Rather, we must concentrate on some specific feature of communities, and preferably one that can both be measured and experimentally manipulated, and then ask whether, with respect to that feature, a community is stable, and if so, why. In many standard ecology texts, you'll see long lists of community features or properties. However, two stand out from the rest, particularly if you're prepared to accept the view of a community as an 'interaction web', as discussed in Chapter 3. At its simplest, this view envisages communities in abstract form as 'dots and arrows', the former being the species, the latter their interactions. If we view communities in this way, they have two key features – the number of species and of interactions respectively. It's possible to imagine two communities which differ radically in the number of constituent species; but it's also possible to imagine two communities with an identical number of species, but differing markedly in how connected-up they are, one having many more interactions than the other.

The first of these features is referred to as species richness (or sometimes diversity), the second as connectance. I'm going to concentrate on richness here, for the following reasons. First, it's easier to measure. To estimate richness you merely need to know who's there; for connectance you also need to know who eats whom. Second, there's some ambiguity in connectance, because while I've been pushing the concept of interaction webs in this book, it's commoner elsewhere to talk of food webs. Should connectance refer only to the degree of 'connected-upness' of *feeding* interactions, or should it be more comprehensive? Different viewpoints lead, unfortunately, to different answers. Third, richness is readily manipulable while connectance is not. You can remove or add species to a system, and observe whether it returns to its pre-perturbation level of richness. But making or breaking feeding interactions among the species that naturally comprise a community is another matter entirely. Try stopping lions from eating zebras! Finally, there is a well-known global pattern in species richness (see Chapter 14) which is crying out for explanation, while we lack sufficient data at present to be able to describe, with equal confidence, global patterns in connectance.

Having decided, then, to focus on species richness as our central community variable, we can formulate a fairly concrete, albeit restricted, concept of community stability, and one which is clearly analogous to the concept of population stability that we've employed through the last few chapters. If perturbations away from the prevailing species richness lead to returns towards the pre-perturbation level, then the system is stable. If they don't, then it isn't.

One thing that such an approach omits is information on the *identities* of the species present. It is possible to imagine pronounced stability in species richness coupled with a great difference in the actual species present before and after a perturbation. If we begin to ask about species identities as well as species richness, we open up a whole additional sphere of enquiry. In particular, the question emerges of whether certain groups of species generally 'go together'. For example, in a patch of woodland surrounded by grassland, do the species of ground flora and invertebrates found in one extend also to the other, or are completely different sets of species found in the two communities? I'm not convinced that it's useful, at present, to pursue such questions, especially since some species are very specific to particular types of habitat, while others are much less discerning. So there can be no neat answer to this question. I will therefore generally avoid the issue of species identities, except when we deal in a later chapter with the concept of 'ecological equivalents'.

Are More Complex Communities More Stable?

There is often said to be a 'conventional wisdom' in ecology, that complexity begets stability. Since we're concentrating, here, on one central strand in the complexity of a system – namely its species richness – the so-called conventional wisdom roughly translates as 'more species-rich communities are more stable.' From one viewpoint, this proposal appears to make sense. If we compare a rich temperate community (say a patch of mixed deciduous forest) with an equal area of impoverished temperate community (say an agricultural monoculture, like a barley field), it's clear that, left to their own devices, one is much more transient than the other. Yet to claim that such a comparison reveals some magic link between complexity and stability is premature, to say the least.

One problem rendering the comparison an unfair one is that it involves on the one hand native inhabitants of a given area (e.g. oak) but on the other hand species that are not only alien to the area concerned, but are indeed so modified by man that they are no longer really 'natives' of anywhere. Clearly, a species-rich collection of artificially modified cereal crops would be almost as transient as its monoculture equivalent, so it isn't just the difference in richness that's involved here.

Having said that, it is nevertheless true that monocultures of *native* species are extremely transient. Anyone who has tried to maintain a top-quality lawn knows this only too well. A monoculture of a native 'fine' grass species – or even a mixed culture of a handful of such species – is extremely prone to being invaded by 'coarse' grasses as well as by broad-leafed species such as buttercups, daisies, clovers and plantains, and indeed by lower plants such as mosses and lichens as well. Does this mean that we can resurrect the link between complexity and stability?

No, it certainly doesn't. Indeed, I would propose that the very idea that any such link exists is based on a misunderstanding of the concept of stability. As we know by now, stability involves the return to an equilibrium value (or band of values) following a perturbation away from it. Inevitably, then, a *stable* system will change rapidly when it's far from its equilibrium, more slowly when in the vicinity of that equilibrium, and (in theory) not at all when it's at the equilibrium. That is, stability and stasis can only be equated when the equilibrium state and the actual state coincide. What we are observing, when we note the easy invadability of low-richness systems, is stability, not lack of it. The low-richness community is transient only because the species richness is heading rapidly back towards equilibrium. Furthermore, agricultural systems only show us downward perturbations in richness. We simply don't observe the 'instability' associated with an overly-high number of species. But if experiments of that kind (i.e. 'addition' experiments) were carried out, we would see a rapid *descent* to equilibrial species richness. So in other words the 'conventional wisdom' is a nonsense based on a confusion of stasis and stability, and on observation of the consequences of downward perturbations but not of upward ones.

There is, however, one caveat. Since there are many alternative measures of both the complexity of an ecological community (e.g. connectance) and its balance or broad-sense stability (e.g. persistence, robustness and others[27]), perhaps, given a suitable choice of measures, a simple relationship between complexity and stability can be salvaged. If so, fine: but it's up to those attempting such a salvage to spell out precisely what measures they are using and why the relationship that they propose actually takes place. If they can't do so, we should give them short shrift.

Determinants of Diversity

The alternative approach, which I'll adopt for most of the remainder of this chapter, is to investigate what mechanisms control the number of species that a given community can 'hold'. If we can understand how such mechanisms work, then perhaps we'll appreciate more about the extent to which a certain level of species richness is 'stable'. I'll consider three such mechanisms in this section – niche differences (related to competition), 'switching' (related to predation) and 'cohesion' (related to mutualism).

Before we discuss these individually, it's useful to take an overview and, in a sense, state the obvious. The total species richness of a community has both horizontal and vertical components. The former concerns how tightly 'packed' species can be within a trophic level, the latter how many levels there can be in the system as a whole. By multiplying these two components we arrive at the total richness: a community with 5 trophic levels and an average of 20 species per level will have 100 species overall. (Even in irregular webs, where no clear levels can be recognized – see Figure 6 – the horizontal and vertical components of species richness can still be conceptually separated.) Most current ecological debate concerns within-level species packing. It's generally accepted (though not without some dissent[28]) that energetic constraints limit the number of levels. So we concentrate instead on what controls the number of autotrophs, the number of herbivores, and so on.

One final point before we discuss our three 'determinants of diversity' individually. There is a useful distinction to be made between local communities and what is known as the 'regional pool' of species. For example, consider types of community that frequently occur in patches in the British countryside – bracken, deciduous woods, or ponds. In all of these cases, the number of species actually present in one particular patch – i.e. in a given local community – is less than the number of species capable of living in such community-types and found within the overall region concerned (in this case the island of Great Britain). While the difference may be partly explained by 'chance' factors (such as lack of migration of a particular regionally-available species to the precise location of a local community), it also has a strong deterministic component. That is, certain forces are at work allowing into a local community some members of the regional pool and excluding others. These are the 'determinants of diversity' to which we now turn.

(a) Limiting similarity and guild structure

We have already examined the mechanics of limiting similarity (in Chapter 8), but with the exception of MacArthur's warblers the case-studies we looked at involved only a *pair* of competing species. In nature, the number of species comprising a 'competitive group' is variable from two upwards. It's useful to have a name for groups of species which use the same sort of resources in the same general way and which may therefore compete – such groups are known as guilds. Examples include large grazing herbivores, bracken-feeding insects, canopy-layer trees, and birds of prey. It must be emphasized that membership of a guild does not automatically imply competition with other guild members. This depends on whether, in the guild concerned, resources are actually limiting; but we'll assume, for the purposes of this section, that we're dealing with those guilds in which resources *are* limiting – i.e. competitive guilds. Nor does guild membership automatically imply taxonomic similarity, though as we've seen there is a tendency for those species that are closely related in their niches to be closely related from a genealogical viewpoint too. Applying the limiting similarity concept to the structure of competitive guilds in nature, the general idea that emerges is that species should be regularly 'spaced out' along resource axes. For example, in a seed-eating guild, where an obvious choice of resource axis is seed size, we would expect to find a small-seed specialist, a middling-size-seed specialist and a large-seed specialist (or a correspondingly larger number of specialists, if the resource axis is longer). Each would be separated from its neighbour by a similar number of resource-axis units, reflecting the level of limiting similarity prevailing in the community concerned. Species attempting to invade from the regional pool may be unsuccessful in squeezing themselves between already-resident members of the guild, once the relevant resource axis is fully utilized. Thus limiting similarity is clearly capable, in principle, of acting as a major determinant of diversity.

Is there any evidence that it actually does so? That is, can we find evidence of regularity of niche distribution in nature? There are difficulties in answering these questions because it's usually very hard to determine whether resources are limiting in general and, if so, then which ones in particular. Even if such matters can be resolved, there are many variables by which any particular set of resources could be described. For the seed example, hardness, shape and nutritional content provide alternatives to the simple measure of size. Furthermore, quantifying species' niches, even for a single resource dimension, is by no means an easy task in a natural community, especially in the face of seasonality, feeding differences between sexes and age-groups, variation in the relative abundances of different resources, and so on.

Because of these and other problems, three indirect ways of looking at niche differences and the limiting similarity question have been adopted. The first is to make use of the relationship between niche and microhabitat, the latter term referring to that portion of the overall habitat in which members of a particular species tend to be found. In a seed-eating guild, large seeds will tend to come from different plant species than smaller ones, so the large-seed specialist should be found (at least when feeding) in the vicinity of different plants to the other members of its guild. Equally, in a guild of insectivores, members feeding on different species of insect should be found in different places. This is one possible interpretation of the microhabitat differences exhibited by MacArthur's warblers, which were illustrated in Figure 14. However, we have to acknowledge the tentativeness of any 'conclusions' derived from such studies, because of the lack of *direct* information both on diets, and on the degree to which the resources concerned are limiting.

The second indirect way of looking at niches is through morphology. For example, if closely-related species of birds have different beak sizes or mammals different jaw sizes, then these may reflect statistical differences in the ranges of prey sizes consumed. If so, regularity in the spacing of guild members along a morphological axis such as beak size may imply regularity of niche spacing, and hence limiting similarity. A famous example of this kind of reasoning concerns the small, medium and large ground finches of the genus *Geospiza* ('Darwin's finches'), as illustrated in Figure 22. Again, however, we have to be very cautious in accepting such indirect information. In particular, while populations of birds may

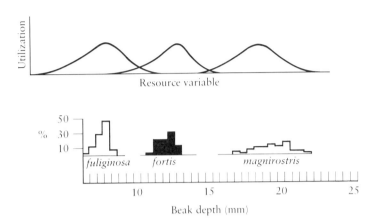

FIGURE 22   Beak depth distributions (*bottom*) and presumed resource utilization curves (*top*) for three species of Darwin's finches on the Galapagos Islands.

indeed be characterized by both beak-size distributions and prey-size distributions, an individual bird, with a particular, fixed, beak size is not restricted to a single size of food item. So the correspondence between morphology and feeding is not a straightforward one; and the many niche studies based on morphological measurements have not taken as full account of this fact as they might have.

The third way of getting at the potential role of limiting similarity indirectly is to make use of the fact that congeners are typically (but not necessarily) closely related in ecological characteristics. Given such a relationship, local communities should exhibit a lower number of species per genus (i.e. have a lower species/genus ratio) than regional pools, because of the competitive exclusion, at a local level, of species that are 'too close', in niche terms, to others in the same community. At first, it was thought that this was so. For example, an early study on British flowering plants showed a species/genus ratio of 2.61 for the regional pool, but values of around 1.2 for local communities. However, C. B. Williams[29] showed that there was an automatic tendency for small samples of species drawn at random from the regional pool to have lower species/genus ratios – in the extreme case of a sample of a single species, it must have a ratio of 1! When the actual ratios of local communities were compared with those that the built-in statistical bias inherent in sampling would produce, local communities were found to have consistently *more* species per genus than expected, not fewer.

This result should not, as Williams pointed out, be taken as evidence that competition-related processes such as niche differences and limiting similarity are unimportant in natural communities. Rather, what it tells us is that any competitive influences tending to reduce species/genus ratios in local communities are overridden, at least in the group concerned, by the effects of shared adaptations of congeners, which tend to increase the ratios. It remains undetermined whether both effects are large and important or small and insignificant – all we know is that whatever their individual importance, one is slightly more effective than the other.

Williams' use of 'pretend' communities, in the sense of randomly chosen collections of species with which to compare actual communities, is an early example of what has now become known as a 'null model'. The basic idea – much developed over the last decade or so by the American ecologists Ed Connor and Dan Simberloff – is that we cannot infer the action of a community-structuring agent such as limiting similarity from examination of a set of data on an actual community alone. Rather, we should form some idea of how a community would look in the absence of the structuring agent concerned (the null model), and resist drawing any conclusions about nature until we have compared it with the null model and seen whether there is a difference – i.e. whether there is any real structure to be explained in the first place. The urge for this kind of caution is a sensible one, despite the difficulties that often exist in devising satisfactory null models in particular cases.

To conclude, then, it's a plausible proposal that wherever competitive guilds are found in nature, limiting similarity is one of their structuring agents. However, a belief in the importance of limiting similarity is at present just that – an act of faith if you like – because enough conclusive studies simply haven't yet been conducted. I suspect that when they have been, we'll end up accepting, rather than rejecting, the limiting similarity hypothesis. However, in *non*-competitive guilds we must look elsewhere for determinants of diversity, which leads to our next topic.

(b) Predator switching and web structure

In cases where resources are not limiting, exploitative competition does not occur, and consequently the idea of limiting similarity is inapplicable. In such cases, it's natural to turn from resources to 'enemies' in our search for an understanding of how the diversity, or richness, of a community is determined. In this context, a kind of predator behaviour known as switching may be particularly important. This wasn't considered in Chapter 9 because we concentrated there on single-prey systems, while switching is only possible, as we'll see shortly, when a predator consumes two or more kinds of prey. In nature, it's usually 'more' but I'll start with the special case of a predator consuming just two species of prey, so as to get across the basic idea of switching in the simplest possible system. Having done that, we'll then look at more complex situations.

Consider how a predator (a bird, say) might respond to variation in the relative frequency of two kinds of prey (for example two species of snail with differently coloured shells). The easiest way to think about this is in the context of an experiment conducted in a series of large greenhouses each with a 'captive' bird, a specially planted ground flora and an introduced snail colony. Different greenhouses have the same overall density of snails but vary in the relative frequencies of the two species (call them A and B) from 5 per cent of the total colony belonging to species A, to 95 per cent. Suppose we monitor the numbers eaten of each species in each greenhouse over an appropriate time-period – perhaps a week.

In such an experiment, three kinds of outcome may be envisaged, and these are illustrated in Figure 23, where the frequency of species A in the predator's diet is plotted against its frequency in the overall snail colony. The simplest kind of behaviour is 'random predation', indicated by the diagonal line. Here, the predator simply eats prey in whatever proportions they are available. A

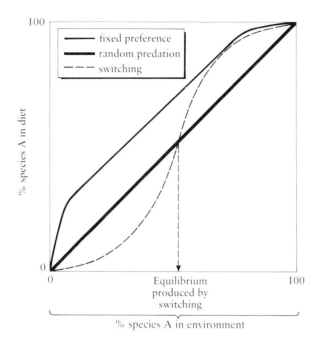

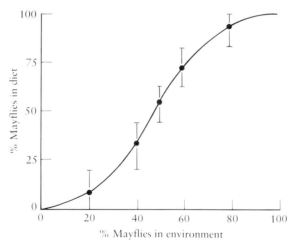

FIGURE 23   Patterns of predation (*top*), including switching, and an example of switching (*bottom*) in the water beetle *Notonecta*.

predator behaving in this way would have no effect on the species composition of the prey colony. The second possible outcome may be referred to as fixed-preference predation (the line parallel to the diagonal but curling in towards it at the ends). Here, the predator consistently 'prefers' one species of prey. I'm using this term in a very broad sense – species A may be 'preferred' because its members are more visible, tastier, larger, more mobile or whatever. Preference is taken, here, to be a measure of what happens, not what's going on in the predator's mind. Obviously, fixed-preference predation could occur in various degrees and in either direction. In any case, its pattern curls in at the ends simply because if there is 0 per cent of species A in the snail colony, there must be 0 per cent of A in the diet. A predator with a fixed preference will tend to increase the frequency of one species of prey (the less preferred one) at the expense of the other – that is, it has a directional effect on the species composition of the prey colony.

Now we come to the interesting pattern – variable-preference predation, or 'switching' (the S-shaped curve). Here, the predator switches its preference in response to the prevailing relative abundances, always concentrating on the commoner type. The pattern of switching illustrated will stabilize the species composition of the prey colony at 50/50, because if either species exceeds 50 per cent its frequency will be depressed by the predator. Of course, this is only one of many possible switching patterns, and the equilibrium could fall at any prey–species composition – its position simply depends on where the 'S' crosses the diagonal, which will doubtless differ from one situation to another. But the important thing is that *all* patterns of switching (though not all patterns of variable-preference predation) will act in a stabilizing manner.

Numerous experiments of this kind have now been carried out both in the laboratory and in semi-natural situations (like our hypothetical greenhouses); and it's clear from them that many predators do, at least under certain conditions, switch. From a behavioural viewpoint, it would be of interest to know why this is so. Do predators form 'searching images' for particular kinds of prey? Is switching an efficient way to hunt? All sorts of questions suggest themselves. But from a community-structure viewpoint, what is more important is whether, in *natural* communities, there is any evidence that switching occurs and thus has a stabilizing effect.

A kind of field experiment which strongly suggests that it does is the 'top predator removal experiment'. As its title suggests, this involves the experimenter removing, or severely depleting the numbers of, the top predator of a particular food web or sub-web, and monitoring what happens to the species composition of the 'prey community' – i.e. the rest of the web 'beneath' the predator concerned. One well-known experiment of this kind was conducted by the American ecologist Robert Paine[30] on a rocky shore community on the Pacific Coast of North America, where the top predator was a starfish (*Pisaster ochraceus*). In areas where starfish were removed, a previously diverse prey community of many species of molluscs and crustaceans collapsed towards a monoculture of mussels (*Mytilus californianus*). In control areas where no starfish were removed, the diverse prey community persisted.

This result is consistent with the hypothesis that (1) in the absence of *Pisaster*, prey species compete for space on the rocks; (2) the competition is intrinsically unstable with *Mytilus* being the dominant competitor under most environmental conditions, but that (3) when *Pisaster* is present it balances a potentially unstable system, by concentrating disproportionately for most of the time on what would otherwise be the superior competitor, switching to other prey only if *Mytilus* becomes temporarily depressed in numbers.

Of course, other interpretations can be made of Paine's results, but at least we can say that they are consistent with a hypothesis of predator-switching stabilizing an otherwise unstable system. Results from predator removals in entirely different kinds of system are also consistent with this hypothesis – e.g. removal of herbivores in terrestrial grassland systems causing a decrease in plant species richness and a takeover by one or a few dominant species.

When we discussed competition in Chapter 8, an important process called frequency dependence emerged. This process, an interspecific equivalent of density dependence, involves each species in a pair or group of species being 'boosted' when at low frequency relative to the others, and suppressed when at high frequency. In the context of competition, the mechanism producing this effect is a niche difference: a temporary excess in species A's numbers will deplete its own resources more than those of species B (with a partially overlapping niche). Predator-switching has the same effect – frequency dependence – but for a different reason. A species whose population expands to very large numbers now suffers not because of resource depletion but because of attracting the attention of predators. I'll expand shortly on the *general* role of frequency dependence: arguably it is as much at the core of community stability as density dependence is at the core of population stability.

(c) Mutualism – ecological glue

I mentioned earlier that ecologists have invested much more effort in investigating those interactions that would at first sight appear to threaten ecological stability – competition and predation – than in studying mutualistic interactions, where each species benefits the other and consequently it isn't difficult to see how the interactants persist together. This bias has paid off in the sense that stabilizing forces such as niche differences and switching have been revealed in competitive and predatory systems, enabling us to see how apparent threats to stability may actually *help* in the stabilizing process. However, our bias also carries with it a danger that we will overlook the role of mutualism as an agent of community stability, which would be decidedly unwise.

I should emphasize first of all that mutualism is exceedingly common. Unfortunately, the opposite view is fostered by people equating mutualism with unusual, 'one-off' associations, like that between leguminous plants and the nitrogen-fixing bacteria that they harbour. Such a false equation is, I think, unwittingly encouraged by the examples given in some elementary ecological texts. In reality, however, there are many diverse kinds of mutualistic interactions. One is the pollination of flowering plants by insects. Another is seed dispersal by birds. A third is the gut bacteria of ruminant mammals, which gain a home and allow their host to digest cellulose, which it would on its own be unable to do. A fourth is found in lichens. I earlier referred to these as 'lower plants' but in fact they are mutualistic associations of algae and fungi. Many further examples could be given, but I think that by now the point has been made: mutualism is very common in nature, and deserves to be ranked along with competition and predation, rather than as something 'unusual'.

That being so, what effect will mutualism have on the diversity and stability of communities? The most obvious answer is that it will have an enriching effect, because when one of a pair of mutualistically interacting species exists in a community, its 'partner' will tend to do so too. Also, it will have a stabilizing effect, because an increase in the population of one of the interactants will tend to boost the other: more flowers means more food for pollinating bees, while more bees means more efficient pollination. That is, mutualism is *intrinsically* frequency-dependent. We don't have to dissect its population dynamics in detail to arrive at this conclusion, the way we did for the more cryptically stabilizing processes of competition and predation.

An interesting area for future research will be the interplay between mutualism and the other types of interaction. For example, do different species of flowering plant compete for a supply of pollinators or, conversely, do different species of pollinator compete for flowers? Consideration of such possibilities introduces an interesting complexity into the relationship between mutualism and stability. If a leguminous plant is competitively displaced by a non-leguminous one, the overall richness of the system will go down. More generally, because mutualistic species go around in pairs or groups, their invasion of, or extinction from, a local community will have a greater effect than in the case of 'go-it-alone' species ; that is, they will make species richness jump up or down by more than single units.

Finally, a quick comment on species identities. Although I said that I'd leave that topic to a later chapter, mutualism has such an obvious effect here that I can't pass over it. Mutualisms tend to be quite species-specific – though how much so varies a lot from one example to another. Thus they represent an agent of cohesion in the natural world, leading as they do to pairs and groups of species that 'go around together'. In some cases, we don't really notice this, because the mutualism is so tight that we tend to think of the whole thing as a single species – as in the case of lichens. But in other cases – e.g. the ants that live inside the thorns of some acacias and help protect them from herbivores – the association is less tight, and the species' separate identities more obvious. In such cases, we can more clearly see the ecologically cohesive effect of mutualistic interactions.

Communities as Self-Levelling Systems

I now want to develop a view of a community as, believe it or not, a stack of rubber sheets. Each sheet represents a trophic level, and each species at a particular level is represented by a circle or ellipse drawn on the appropriate sheet (see Figure 24). In this view, a population whose numbers are at their equilibrium is indicated by the rubber sheet being flat and undistorted in the vicinity of the circle representing that species. An upward distortion, or dome, in the sheet represents excessive numbers in the population concerned, while a downward distortion, or pit, represents population depletion. Another aspect of this picture is that each circle in a particular sheet can be 'mapped' to particular circles in the sheets above and below it, according to which species of predator and resource it interacts with.

Let's consider what happens when, starting with a community at equilibrium (all sheets flat), one particular population expands dramatically, say because of a sudden influx of immigrants from a neighbouring area. This is shown happening in one of the species in trophic level 2 in Figure 24. What will the consequences of this population expansion be for the rest of the community? Well, the greatest impact will be on those populations with which the 'expanded' one directly interacts. Its resources will be depleted (pits in lowest sheet), while its predator will increase in numbers (dome in upper sheet). Both of these effects will then serve to neutralize the event that set them off in the first place – that is, the expansion in number of one of the populations in the middle sheet. Thus the nature of a community as a self-levelling system where

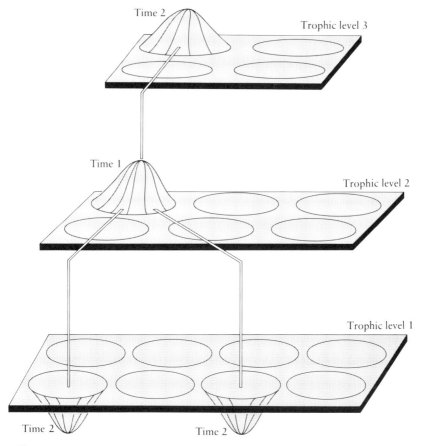

FIGURE 24   'Rubber sheet model' of communities, devised to illustrate their
inherent self-levelling properties.

negative feedback effects act to negate any departure from the
equilibrium state is readily appreciated.

Of course, we saw at an early stage in this book that the concept
of the trophic level is, for many natural communities, a rather
dubious one. But this doesn't affect our argument. An alternative
way to picture the self-levelling process is to consider a community
as an interaction web, with each species in the web being
represented by a ball of variable size. Population expansion can
now be represented by a swelling of the ball, depletion by

shrinkage. In such a scheme, it's easy to picture a direct equivalent of the neutralization of a population expansion such as we observed using the rubber sheet model.

As well as fuzziness of trophic levels, real systems add all sorts of other complexities to our neat self-levelling picture. Foremost among these is the fact that in nature it's almost certain that not only does the *actual* population size of a species keep changing, but also the *equilibrium* that the actual numbers can be thought of as seeking will itself probably change, even on a restricted timescale of a handful of generations. This merely reinforces the point with which this chapter started: natural communities are notoriously difficult to deal with, while conceptual models – whether involving sheets and balls or mathematical equations – are a lot simpler.

How does the self-levelling picture connect with our main agents of community stability, such as niche differences and predator-switching, and the density and frequency-dependent forces to which they give rise? The answer is, of course, that it is precisely these forces which do the levelling, though it would take a much more sophisticated picture than Figure 24 to do justice to the full complexity of the interactions involved in any one natural community. There are, however, two rather different ways in which density and frequency dependence operate in nature, and I think it's important to distinguish these.

The first is reasonably analogous to laboratory microcosms, greenhouse experiments, and other human-engineered communities. What I'm referring to here is the operation of these stabilizing forces *within closed systems*. In such systems, all stabilizing forces act through density-dependent natality and mortality. Although natural systems are intrinsically 'open', it's probably true that in most of them a large component of their stability operates through birth and death rates rather than being dependent on the migration of organisms to/from neighbouring communities.

Having said that, the fact that most natural communities are indeed open systems means that there is a second way in which stabilizing processes can operate in them, namely through density-dependent migration. For example, in our rubber sheet model (Figure 24), one reason why the predator begins to 'dome' after the population explosion in its prey is that its birth rate goes up and its death rate down. But another reason is what I have seen called the 'pantry effect' – the attraction of predators to localities where prey density is unusually high. I would guess that in nature it's usual for density-dependent responses to have both within-system and migrational components, with their relative contributions varying considerably from one community to another.

A colleague who is an evolutionary biologist rather than an ecologist once asked me whether there was any one central process in community ecology equivalent to natural selection in evolution. My immediate reply was density dependence. It may stem from different sources in different systems, and it may on some occasions work largely through birth and death rates, on others through migration. But the universality of its occurrence, despite the differences in detail from place to place and from time to time, is now beyond doubt, as is its intrinsically stabilizing effect.

On reflection, my only qualm about that immediate answer was that perhaps I should really have said 'frequency dependence'. After all, what is critically important in nature is not precisely how many individuals there are in a given population, but rather how their numbers compare to the numbers of the same species in neighbouring localities, and to the numbers of resources, competitors, predators and mutualists. Of course, what constitutes the critical frequency – that is, in relation to what other populations the numbers in the one under study should be compared – will doubtless vary from one situation to another. But in general terms, it is usually the ratio of one population size to another (or others) that is crucial, not just the absolute numbers of 'population X' in a vacuum. And consequently we should think of frequency dependence broadly defined, not narrow-sense density dependence, as the foremost process in community stability.

The exercise with which this section began involved consideration of a sequence of changes (in the rubber sheet model) whose starting point was the equilibrium state. In nature, no newly colonized area starts from anywhere near the equilibrium. The process of getting from a few colonists to the 'climax' community is our next port of call. Are successional communities also self-levelling or do they behave in a fundamentally different way? The following chapter takes an in-depth look at this question.

11

From Rock to Rainforest

Suppose that, sometime tomorrow, an undersea volcano were to erupt, disgorging millions of tons of lava. If the eruption goes on for long enough, the surface of the volcano will rise above sea level, and it will become an island. Initially, it will simply be a lump of volcanic rock devoid of life. However, if the island is located in a moist tropical region it will probably, within a thousand years, be covered in rainforest. The process of getting from a virtually abiotic environment to a climax community such as a rainforest is called ecological succession.

Of course, like all other ecological processes, succession is best regarded not as a fixed sequence of events with clearly defined properties but rather as a set of variations on a central theme. In succession, the central theme is a directional trend from a simple, low-diversity beginning to a complex, high-diversity climax. Precisely what is the starting point will vary from place to place, as will the nature of the climax community that is ultimately arrived at. The length of time required to get from start to finish, and the precise pattern of events that takes place over this timespan, would also be expected to vary. Nevertheless, we shouldn't allow such locality-specific details to obscure the very real similarities of successional systems generally.

An important distinction is that between 'primary' and 'secondary' succession. A new volcanic island constitutes a system of primary succession because the substrate that is being colonized did not previously support any other ecological community. Another example of primary succession is the colonization of an area of bare sand. Secondary succession, on the other hand, occurs when the starting point is an area of land (or water) that already has, or has had in the past, a biota of some sort. Two examples of this kind of succession, which I briefly mentioned in the previous chapter, are the development of forest from abandoned farmland, and the gradual re-emergence of a climax community following its destruction by fire.

These two types of succession, primary and secondary, differ in two quite important ways. First, dispersal is less of a problem in secondary succession, because many of the organisms which will later be involved in the successional sequence are already present (some as seeds) when the sequence starts. Second, some of the directionality of the process is, in secondary succession, due merely to the longer development times of 'later' species. We know that trees can't grow on unweathered bare rock, so that in a primary succession the appearance of trees towards the end of the sequence *must* be due at least in part to earlier species modifying the environment in such a way as to make it habitable to trees. But in a secondary succession where soil is present right from the start, it *could* be argued that the absence of mature trees early in the sequence and their dominance later on is simply a result of the fact that trees take decades to develop while grasses (for example) can appear in a matter of days. Having said that, it's most unlikely that any secondary succession can be explained *entirely* on the basis of variation in species' development times. The 'typical' secondary succession involves an element of this, but also an element of the progressive environmental modification found in its primary equivalent.

Because the appearance of novel abiotic environments such as volcanic islands is rare, most studies of succession have been based on secondary systems, and one popular sort of case-study has been the abandoned agricultural field. The transition from a starting point such as short, grazed grass to mixed deciduous forest (assuming a temperate location) takes something of the order of a hundred or so years (see Figure 25), in contrast to the thousand or

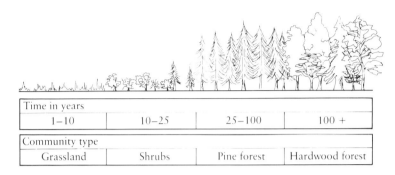

Time in years			
1–10	10–25	25–100	100 +

Community type			
Grassland	Shrubs	Pine forest	Hardwood forest

FIGURE 25   Approximate times to various stages in a terrestrial succession in the temperate zone.

so that may be required to get from pristine rock to the appropriate climax in a 'pure' primary system. As can be seen from the figure, there is a general increase in the complexity of the vegetation as we move from earlier to later stages in the sequence.

Two things should be admitted at this stage. First, as Figure 25 reflects, much work on succession has been botanically orientated. This is simply because the plant cover of a region is relatively obvious and static compared to the fauna, much of which (e.g. the soil invertebrate component) is both cryptic and mobile. So it's a lot easier to observe a temporal sequence in the vegetation than in its associated animals. Having said that, it's true that a directional change in the flora over a period of time is usually associated with a directional change in the fauna.

The second admission concerns our tendency to infer temporal sequences from observation of changes from place to place at a particular moment in time. For example, we may look at a transect of samples across a stretch of sand dunes, starting at the seaward side with 'embryo dunes' and ending, some distance inland, with 'stabilized dunes' where the familiar marram grass is vastly outnumbered by plants typical of non-coastal terrestrial communities. It's a reasonable mental jump to conclude that the stabilized dunes were once embryo dunes and that the different communities observed at different points on our transect represent localities that differ in age and hence in their position in the successional sequence. But we mustn't forget that it *is* a mental jump, and that evidence gleaned about temporal processes from observations made at a particular point in time is necessarily indirect. The use of such indirect evidence is of course simply a pragmatic solution to the problem of studying a process which takes longer than one's own lifetime. This problem is even more pronounced when processes operating on an evolutionary timescale are the focus of attention, as we'll see in the next two chapters.

Occasionally, we get the opportunity to observe at least part of a successional sequence as it actually happens, usually on a rather small spatial scale. An example is a newly built dry-stone wall separating upland fields – not that many are constructed nowadays, given the relative ease of erection of barbed-wire fences and the like. But if you *do* get the opportunity to observe such a wall – or any other structure made out of recently quarried stone and then left thoroughly alone by man – you'll observe a fairly orderly sequence of events. Lichens are usually the first to colonize. As they grow, accumulate wind-blown debris, partially decompose and physically weaken the initially impermeable surface of the rock, mosses begin to appear. After a period of time, grasses and a few broad-leafed plants are observed growing on the wall where only lichen used to be found.

Unfortunately, the physical nature of the wall, with small sub-units and a vertical layout, is not conducive to the progress of succession beyond this stage. But if we were dealing not with a few stones piled on top of each other, but with a flat expanse of rock in a reasonably protected locality, then, if we lived long enough, we should see the grassy stage give way to scrub and eventually to forest.

Succession and Species Richness

As a general rule, during succession the number of species present in the community tends to increase. This change is seen most dramatically in primary successions, where the start-up richness is zero, and particularly those in tropical regions where the climax community may contain thousands of species. The trend is less marked both in temperate and polar locations, where the terminal diversity is lower, and in secondary successions generally since there the initial diversity is higher. Nevertheless, comparing the start with the finish, virtually all successional systems exhibit a net increase in species richness. In cases where we're able to track the number of constituent species in the community *throughout* succession, we sometimes find that the pattern of change in this variable is complex. In particular, it's quite possible that we may observe decreases in species richness during particular successional phases. One reasonably common pattern is for a marked and continuous increase for the majority of the time, from initial 'pioneering' stages through to about the penultimate stage, followed by a final (and relatively slight) decline. One example of this kind of pattern is given in Figure 26.[31]

Why should such a pattern be found? There are at least two possible explanations. The first is that in a sense the terminal decline is illusory. If you picture succession as a sequence of different, but temporally overlapping, community types, then in the middle of the sequence the total species richness will be inflated

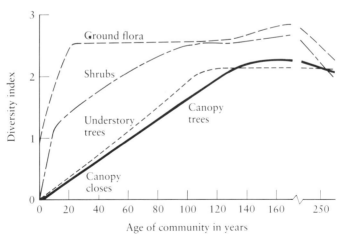

FIGURE 26 An example of changes in species diversity during succession. Note early major increase and later, relatively minor, decline. Data are from a field-to-forest succession in Georgia.

146

by the combination of two partially different 'species lists', one for the community that is currently on the way out and one for that in the ascendancy. At the finish, such compounding effects will disappear, with a consequent decline in species richness that is, at least from some viewpoints, illusory.

The other possible explanation for terminal declines in diversity is that they are real, and are caused by some special features of the particular kind of climax community found in particular places. The most obvious example of this sort of thing is the coniferous forest climax of northern temperate regions such as much of Canada, Scandinavia and Siberia. Here, the success of the canopy layer (itself rather species-poor) in preventing much of the incident sunlight from penetrating through to ground level has an obviously depressant effect on the species richness of the ground flora and its associated fauna. This is in contrast to the greater richness at ground level in the earlier stages of succession from which the forest climax is derived.

While it's important to clarify as far as possible the exact pattern of change in species richness over time in successional systems, and to try to explain any downward 'blips' in a generally upward trend, it's even more important that we don't let such blips distract our attention from the general tendency for species richness to rise. What we need to ask ourselves next are questions of mechanics. We need to know both *how* some species replace others (competitive exclusion?; extinction and recolonization?) and also *why* richness tends generally to increase. That is, we must move on from the observational stage wherein we merely note that these things happen, to the more investigative stage of trying to analyse their causal connections.

The Mechanics of Species Replacement

It's probably true that all the kinds of interaction between species that occur in climax communities occur also during every stage of the successional process from which the climax arises. It's certainly impossible to have a community lacking in trophic interactions. It's also rather unlikely that early-stage communities lack competition. Competitive interactions may be less common in pioneer communities than in climaxes (though we don't know for sure that they are), but they are certainly not absent. However, if we're to explain the very definite directional 'flavour' of succession, which is in a sense the antithesis of the balanced flavour of the climax, we should perhaps be looking not to those interactions prevailing in communities at all stages of their development, but rather to those that are found during the transitional phase leading up to the climax but not (or at least not much) in the climax community itself.

Are there any such interactions? To answer this question, it will help if we have a scheme for classifying interactions – preferably a complete, all-inclusive one. I've presented such a scheme, modified in one respect from the well-known scheme of American ecologist Eugene P. Odum, in Figure 27.[32] Let's now spend a little time working through this scheme.

The basic idea is that if two species interact, each may have a

		Effect of species A on species B		
		Beneficial	Neutral	Detrimental
Effect of species B on species A	Beneficial	++ Mutualism		
	Neutral	+o Commensalism	oo 'Neutralism'	
	Detrimental	+– Contramensalism	–o Amensalism	–– Competition

FIGURE 27   A classification for interactions between the populations of two species.

beneficial, detrimental or neutral effect on the other. For example, competition between species is mutually detrimental since both species get less of the limiting resource than they would have done if the other was absent. Equally, predation is clearly beneficial to the predator and detrimental to the prey, and so on. If we represent the three kinds of effect with the symbols + (beneficial), ○ (neutral) and − (detrimental), then all interactions between species must fall into one of the six (or really five) boxes shown in Figure 27. Of course, if we choose to take into account how the effects are achieved, then many more categories are required. But if we're content to restrict our attention to effects for the sake of simplicity – and leaving open the option of looking at mechanisms later – then we end up with just five kinds of interaction.

These five can be grouped to simplify things further. The (−○) category, called amensalism, can be thought of as just the most extreme case of a continuum of progressively more lopsided cases of competition. So we can lump these two together. Equally the (+○) category, called commensalism, can be thought of as the end of a spectrum of increasingly asymmetric mutualisms, so again we combine the two.

That leaves us with (+−) interactions. Now in most ecological texts, such as Odum's, that category is equated with 'predation' (broadly defined to include parasitism and herbivory). But here we come up against a problem. Predation is an *example* of a (+−) effect but it isn't synonymous with it. When the mechanism of achieving a (+−) effect is one species feeding on another, then of course that's predation. But is it not possible to have (+−) interactions achieved in other ways? Of course it is, because if one species modifies the environment in such a way that it facilitates the invasion of a second, but the invader then acts to suppress its benefactor, then a (+−) interaction is achieved without either species feeding upon the other.

Such interactions are probably common in succession, and I would hazard a guess that they are one of its major driving forces. Yet ironically they are denied a name in the standard classification. To rectify this problem, I recently proposed[33] that we call (+−) interactions 'contramensalism' and that, depending on how the contramensal effect is achieved, we can partition this overall category into trophic contramensalism (predation) and non-trophic contramensalism (for which there is as yet no name).

I'm not suggesting for a moment that non-trophic contramensalism is the only driving force of succession. Other processes are probably also at work.[34] Among them are the demise of species by self-inhibition, and straightforward competitive exclusion comparable to that which we observed in laboratory systems in Chapter 8. Indeed, I suspect that it is no more possible to identify a single cause of the intrinsic directionality of succession than it is to identify a single balancing agent (such as predator-switching) responsible for the stability of the climax. But we can and should attempt both to list the possible mechanisms, and to comment on their relative importance. My own guess (or hypothesis, if you prefer) is that contramensal interactions involving modification of the environment are at the top of the list.

Succession as a Self-Organizing System

While contramensal interactions operating through environmental modification provide a mechanism for species replacement, they don't explain why, as the replacement process proceeds, there is a net increase in diversity. It is to this aspect of succession that I now wish to turn. I'll first consider the question of how we should view successional increases in diversity – for example, are they best seen as part of a self-levelling process? I'll then turn to questions of mechanism – how are increases in diversity brought about, and what ultimately stops the trend in this direction when the climax stage is reached?

With regard to our 'conceptual overview' of the process, there are two alternative possibilities. One is to view succession as the process by which the ecological community of a particular area 'builds itself'. Systems which have this power of constructing themselves as complex entities out of relatively simple beginnings are described as self-organizing. Alternatively we may view succession simply as a special case of the more general category of 'departures from equilibrium' and therefore as part of what in the previous chapter I called the self-levelling process.

It makes a lot of sense to take what may be considered a compromise view and to regard primary succession as self-organizing, and secondary succession as part of the general self-levelling behaviour of the system that is the end-point of the self-organizing process. There are two reasons behind coming to this decision. First, it seems decidedly perverse to regard a recently formed barren volcanic island as a perturbation away from the climax community – e.g. rainforest – that will eventually be found there. Therefore, it doesn't seem appropriate to consider the primary succession that leads up to the initial climax community as part of the self-levelling behaviour of that community, since it doesn't yet exist. But any kind of disturbance (fire, storm damage etc.) that temporarily impoverishes the already formed climax and gives rise to a process of secondary succession may quite legitimately be regarded as initiating a sequence of self-levelling behaviour.

The second reason for regarding primary and secondary succession as different types of process is that the actual species-replacement mechanisms may be different. In primary succession, environmental modification by existing species facilitating the entry of subsequent ones is a necessary, indeed major, part of the process. That is, the facilitation-inhibition kind of contramensalism is crucial. But in secondary systems, where necessities such as soil are already present at the outset, straightforward competitive exclusion may be more important.

Of course, it's always difficult to draw neat distinctions between ecological processes. Rather than there being clearly defined primary and secondary types of successional process, there is really a continuum from 'pure primary' through varying degrees of secondary to very minor bits of self-levelling behaviour that would hardly be regarded as succession at all. Embryo dunes are a lot less abiotic than recently solidified lava, even though they are conventionally seen as a starting point of primary succession. And fires can destroy all the above-ground biota, while temporary conversion of forest to farmland does no such thing. So secondary successions are quite heterogeneous also. The re-invasion of a single species of plant (say) after a local extinction would hardly be considered as an example of secondary succession, while the re-invasion of several species of broad-leafed plants into a grassland system following their removal by selective herbicide probably would qualify for secondary succession status. So we can see a continuum all the way from absence/destruction of whole communities through to extinction and re-invasion of a single species. But having said that, the primary and secondary categories are still very *useful*, even if the line separating them is a bit fuzzy. If we refuse to make any distinctions because they're not perfect, clear-cut ones, then we'll wallow in a mire of heterogeneity and will never make progress at all.

It's now time to turn to the question of the *mechanism* underlying successional increases in species richness. The easiest way to deal with this problem is to use the niche concept, which we've already encountered when discussing competition in Chapter 8. Remembering that a population's niche is some measure of what it does/needs, often in terms of feeding behaviour (animals) or soil nutrient requirements (plants), let me introduce the complementary idea of niche space, which is a measure of what there is to do. That is, niche space characterizes environments, niches the populations that inhabit them.

In secondary succession, there is empty niche space, in the sense of things to do that are not being done – e.g. no canopy-layer trees despite a soil that will support them. So while part of the process involves the competitive replacement of one species by another, some species can invade and establish populations without displacing anything, and therefore the overall species richness increases. In primary succession, something broadly similar happens except that the niche space that becomes 'occupied' is actually brought into being by the successional process rather than being there to begin with.

Why can the trend towards increased species richness not go on indefinitely? Again, recourse to the niche concept provides a possible answer. Because of the limiting similarity criterion, which as we've seen precludes the coexistence of species that are too close in niche terms, there is ultimately a limit to the number of plant species that a given region can 'hold'. This in turn sets a limit on the number of coexisting herbivores, and so on. Energy constraints set an ultimate limit to the number of trophic levels, providing a 'vertical constraint' complementary to the horizontal one set by competition and limiting similarity. When the species richness comes up against these constraints, the system equilibrates, and from then on self-levelling behaviour in response to perturbations will take the form of declines back to equilibrium as well as secondary successional rises to it.

It remains to be seen how our picture of succession will change if the idea of non-resource-regulated populations, and consequent non-competitive guilds, turns out to gain acceptance. One possibility is that our view of plant succession will be unaffected, since in the botanical realm resource limitation may be the norm, but that our view of animal succession (which is much more rudimentary anyhow) will shift away from niche-based explanations.

Species Identities

So far, we've dealt with succession rather anonymously, from the viewpoint of species identities. We've thought of species richness as generally increasing through the successional process, perhaps decreasing slightly in the final stages before climax, and finally stabilizing in the equilibrium zone appropriate to the region concerned. But what *actual species* comprise the 257 (say) that make up the climax community or the three that represent the 'pioneer' community or the 102 that represent a mid-successional stage?

The most important point here is that while there is considerable place-to-place variation in species identities between different local patches of the same community type (e.g. woodland) both in comparisons of their climaxes or in comparisons of a common successional stage, the collections of species are far from random, and there is a high degree of correspondence from one place to another. In other words, there is much more to succession than the attainment of equilibrial species richness. The process also tends towards the attainment of certain species identities making up the overall richness.

Unfortunately, this fact may have been obscured by having been over-stated in the past. In particular, the claim[35] that in any large biogeographical region, succession tended to produce a fixed, repeatable climax community, dependent only on the prevailing climate – the so-called monoclimax theory – was clearly non-sensical. Soil-type, topography and other factors will also be important in determining the direction and end-point of succession, and any given region of moderately large size (say a few thousand square kilometres) would be expected to exhibit several climax communities – perhaps one predominant and others representing relatively localized variations. Having said that, two randomly chosen hectare-sized plots of Canadian coniferous forest destroyed experimentally by fire would probably show a broadly similar succession and climax, as would two hectare-sized plots of rural English lowlands allowed to 'go natural'.

As well as comparing succession and climax in two patches a few kilometres apart, we can also make broader-scale comparisons across biogeographical realms. For example, how would succession of abandoned farmland compare between North America and Europe, assuming that both localities are chosen in areas where temperate deciduous forest is the natural climax state?

The probable outcome of such an 'experiment' is that the succession of *types* of communities would be similar (grassland to mixed herbaceous vegetation to scrub to forest), but most of the *species* involved at each stage would be different, including, at the climax stage, different species of dominant tree. Such a result tells us two things. First, the self-organizing or self-levelling process involved has certain properties, as an ecological system, that transcend the materials of which the system is composed. Second, certain species in one realm can be considered as 'ecological equivalents' of species in another realm. Since ecological equivalents are sometimes only distantly related taxonomically, this suggests that *evolutionary* processes – such as the modification of species by natural selection to fill empty niche space – also operate in a sense independently of their raw materials.

The concept of ecological equivalents is only one of many points at which ecology and evolutionary theory intersect. Consequently, to be a well-rounded ecologist, it's necessary to know something about evolution. In the next two chapters I provide a fairly rapid tour through those aspects of evolution that are most ecologically relevant. This means re-setting our mental clock. We now move from successional changes, which take at most a thousand years, to the history of the ecosphere, which involves a timespan of four billion.

12

THE EVOLVING THEATRE

It's time we returned to our spacecraft. Suppose that, having set off on our 1000-planet sampling mission, we get caught in that dream of sci-fi buffs, the time warp, and find ourselves looking for ecospheres not in today's universe but in the universe of the distant past – a little over four billion years ago to be precise. While it's possible that we would indeed encounter the occasional planet hosting an ecosphere at that time, the earth would not be on our list of 'finds'. The reason is not that there was no earth (there was) or that we omitted it from our sampling programme (we didn't), but rather that, at that time, there were no terrestrial life-forms. In other words, the earth was as barren and unecological as Venus, Mars or anywhere else.

What we're beginning to do, here, is to put a time dimension to the ecosphere. Our initial foray into space in Chapter 1 was designed to broaden our perspective spatially, from earth-bound and parochial to inter-galactic. But it's as easy to be parochial in time as it is in space, and as important to fight temporally blinkered vision as it is to fight its spatial equivalent. The ecosphere had a beginning, and will undoubtedly have an end – whether in tomorrow's nuclear holocaust or in a supernova fireball in the distant future. And the segment of its lifespan that is over is characterized by a definite history which, despite the difficulties in peering back through millions of years unaided by time-warps, we can dimly perceive from diverse indirect lines of evidence including fossils and radiometric dating.

The famous ecologist G. E. Hutchinson wrote a book in 1965 with the marvellous title *The Ecological Theater and the Evolu-*

tionary Play.[36] This conjures up an image of the various species as the 'players', going through their evolutionary motions against a backcloth of a local community and, on a broader scale, the ecosphere as a whole. However, the analogy breaks down because the ecospheric theatre is partially composed of the players, and it's the whole theatre, players and all, that acts out the drama.

There are two main aspects of ecospheric evolution. First, there is evolution in the strict biological sense. Communities have changed over evolutionary time because the species forming the interaction web have changed. There's a world of difference between a green slime composed entirely of unicells and a modern wildlife paradise such as the Serengeti. Second, there have been considerable physical changes in the abiotic 'half' of the ecosphere, ranging from continental drift to the transition from a low-oxygen to a high-oxygen atmosphere. Although these physical changes are not evolution in the usual sense of the word, I'll include them here because they are an integral part of the ecosphere's history and they interact strongly with biological changes in both causal directions. That is to say that biological events can cause physical changes (as in the oxygen build-up), but physical events can also cause biological changes (such as the break-up of a land-mass engendering speciation).

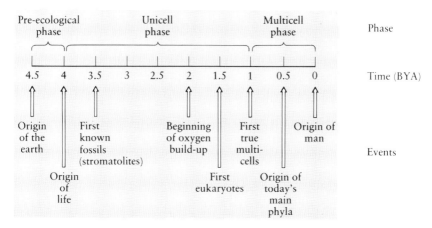

FIGURE 28   Phases in earth history, and some major terrestrial events dated to the nearest half billion years. (BYA = billions of years ago.)

The history of the earth can be roughly divided into three main phases: the pre-ecological phase, the unicell phase and the multi-cell phase. These are shown in Figure 28, which gives dates of certain major events to the nearest half billion years. We'll now take a whirlwind tour through the three phases of earth history.

Pre-ecological and Unicell Phases

The earth is thought to have originated about 4.6 billion years ago (using 'billion' in its now standard sense of a thousand million, or 10^9). Precisely how it originated is outside the scope of this book, and indeed is still being argued about by cosmologists, though it does seem that the whole solar system originated at about the same time.

The pre-ecological phase of earth history, which we can dispense with pretty rapidly, lasted for about the first half billion years. Towards the end of this time, nutrient-rich pools, probably ranging from pond-sized to ocean-sized, were abundant. These contained the so-called primeval soup in which life originated as primitive single-celled creatures superficially similar to our present-day bacteria. Placing this origin – and so the end of the pre-ecological phase – at a definite point in time is impossible since simple unicellular organisms with no hard parts don't fossilize well. Our first known fossils – of large matted structures called stromatolites built up by unicellular creatures – date back to about 3.5 billion years ago, so the origin of life clearly occurred prior to that. But my placing of it at the four billion mark in Figure 26 is somewhat arbitrary, to say the least.

Stromatolites are fascinating because they represent at once both the first known organisms and the first known ecological communities. Most fossils – e.g. trilobites, dinosaurs or *Homo erectus* – are simply individual organisms. But stromatolites represent communities of lots of unicellular and/or filamentous creatures living together and doubtless interacting with each other in various ways, as we know from observations of the few surviving present-day stromatolites found in such places as salt lagoons (see Figure 29).

FIGURE 29   Present-day stromatolites, such as these, closely resemble the
earliest stromatolite communities dating from 3.5 billion years ago.

Strangely, there was what might be called a quiescent phase in
the history of the ecosphere ranging from its inception for about
the next 3 billion years – that is, three-quarters of its total lifespan
to date. During this time, only unicellular and filamentous creatures were found and ecological communities were both visually
undramatic and, as far as we can tell, organizationally simple
compared to their modern counterparts.

The main evolutionary event to occur during this period was the
origin of eukaryotic cells – that is, cells with proper membrane-
bound organelles, such as the nucleus. From an ecological viewpoint, however, this event was of little consequence. Instead of
having bacteria-like cells, some of which photosynthesized, some
of which attacked other cells, and some of which – 'detritovores' –
simply absorbed the nutrients deriving from cell breakdown, there
were now eukaryotic cells doing these same things. A big step for
the cell, but a minor change for community structure, as you might
say.

It's hard to visualize the ecosphere during its first three billion
years, but let's try it anyway. There were probably both terrestrial
and aquatic habitats throughout this time. Many of the terrestrial
ones – those at the drier end of the spectrum – would have been
devoid of life, so there would no doubt have been vast expanses of
bare rock. In places where rocks were kept permanently damp, by
water flowing over them or splashing on to them, and protected by
mist or cloud from ultra-violet radiation, there was probably a

'slime' of unicellular creatures and hence a primitive sort of community. The most 'developed' communities, however, are likely to have been aquatic ones, and it's possible that a primitive kind of 'plankton' inhabited the upper reaches of most water bodies. Add to this the scattered occurrence of stromatolites, and we may well have a reasonably complete mental picture. And the incredible thing about this picture is that it's equally accurate for three, two and one billion years ago. The unicell ecosphere was, at a visual level, a very dull place.

The build-up of oxygen in the earth's atmosphere began about two billion years ago, when only bacteria-like creatures were found. The origin of eukaryotic photosynthesizers will have speeded the process, and by the end of the unicell phase the level of oxygen was approaching 5 per cent of its present value. By half a billion years ago, when the main types of life-form known today had appeared, it had reached a third of its present level, and since a quarter-billion years ago, it has barely changed. But we're beginning to get ahead of ourselves here: let's now consider that most stunning ecological and evolutionary event, the origin of multicellularity.

The Multicell Phase

Until the 1940s it was generally thought that multicellular lifeforms first appeared about half-a-billion years ago, and that these earliest creatures were not unlike our present-day flora and fauna – at least inasmuch as they could be placed within the same major taxonomic groups as extant organisms: for example jellyfish, flatworms, molluscs, insects, and vertebrates. However, in 1946 an Australian geologist, R. C. Sprigg, discovered numerous fossil animals near old lead mines in the Ediacara Hills, about 500 kilometres to the north of Adelaide. These fossils, which became known as the Ediacarian fauna,[37] proved to be one of the most interesting fossil finds ever made. While there is still much debate about what they represent, the prevailing view is that they are the fossilized remains of an evolutionary radiation of multicellular creatures quite different from the ones we now know, and dating from the period between half and one billion years ago, which had previously been thought of only as the tail-end of the unicell phase.

It's not at all clear what kind of communities – in terms of interaction webs – the Ediacarian fauna were involved in. Of course, a multicellular body-form opens up all sorts of possible ecological roles, such as fast-moving predator, which were previously impossible. However, seeing the potentialities and knowing what actually happened are worlds apart. So we shall have to be content with an artist's impression of the strange creatures that inhabited the seas more than half-a-billion years ago (Figure 30), and with the knowledge that there *were* interesting and organized communities then rather than just green slime, even if we're unlikely ever to establish any firm facts about their ecology.

FIGURE 30   Artist's impression of a marine community in the pre-Cambrian, before the origin of today's main multicellular phyla.

The Last Half Billion Years or So

Two of the most dramatic kinds of event in ecospheric history are 'explosive evolution' and 'mass extinction'. It appears that, a little more than half-a-billion years ago, the strange creatures shown in Figure 30 became extinct, and a burst of evolutionary diversification produced most of the major types of organism that we find around us today. At first, these were confined to the sea, but later many invaded the land so that both aquatic and terrestrial communities were established. It's probably true to say that by about a quarter-billion years ago (in the Permian and Triassic Periods, for anyone familiar with such labels) the entire earth's surface was covered with ecological communities that functioned much as our current ones do. All the species will have been different of course. But in a sense we can regard them as ecological equivalents in time.

At about the 'quarter-billion mark', the world had a very different geography from that of today; all the continental land-masses were fused together into a supercontinent known as Pangaea (literally 'all earth'). Between then and now there has been a progressive break-up of this supercontinent, first into 'halves' – called Laurasia and Gondwana – and subsequently into the present-day continents with which we're all familiar. The details of this break-up are shown in Figure 31.

It isn't merely the positions of the continents that shift during evolutionary time. This is in a very real sense only the tip of the iceberg. The outermost layer of the solid earth – the crust – is divided into a number of 'plates' which are in constant, albeit slow, motion in relation to each other. In some places, these giant plates slide past each other, while elsewhere, in a process called subduction, one is pushed down beneath another. Upwelling in mid-ocean ridges produces new plate material which compensates for that lost through subduction. Since the continents are carried on plates, they move around as their respective plates move.

This once-revolutionary theory of the earth's crust (called plate tectonics) is now generally accepted, and forms an important aspect of the history of the ecosphere. It also ties in with

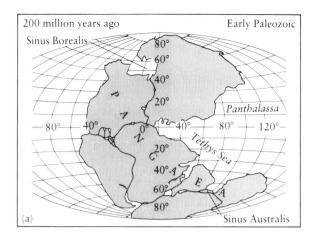

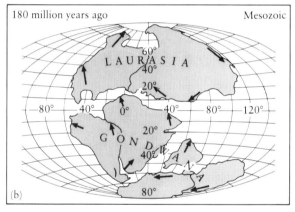

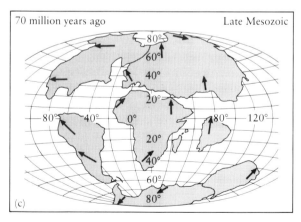

FIGURE 31   Break-up of Pangaea (a), first into Laurasia and Gondwana (b), with further separation (c) yielding continents approaching their 'modern' form.

continental movements other features of the earth's geology such as volcanoes and earthquakes. These tend to be located at plate boundaries, and they reflect the intense geological activity that's going on there. The major plates, and their current locations, are shown in Figure 32.

Where does our now-mobile view of the continents leave Wallace's Realms? In Chapter 2 we noted that the terrestrial component of the ecosphere was divided into six major biogeographical provinces (Figure 1), whose floras and faunas differed considerably from each other. It's now clear that, in terms of earth history, these Realms are a rather recent 'invention', having been in existence only for about 0.1 billion years. That doesn't make them any less important a feature of our present-day world, of course, but it does remind us yet again not to regard even the largest-scale ecological phenomena as being permanent ones. In the ecosphere, all is transitory – though some things more so than others.

From the origin of the main groups of multicellular organisms through all the continental movements just described and up to the present, there has been continuity in the coverage of our planet by life-forms. However, in addition to gradual alteration of the array of life-forms present due to 'ordinary' evolutionary change (which we'll look at in Chapter 13), there have also been a few mass extinctions in which anything up to about 90 per cent of the then-existing species have been destroyed. Particularly major extinction events occurred at 440, 370, 250, 220 and 65 million years ago, the last of these including that most talked-about extinction, the dinosaurs. The reasons for the occurrence of these and other lesser extinction events remain obscure, and all sorts of theories, including those proposing asteroids colliding with the earth, have been put forward. I'm neither qualified nor inclined to attempt an assessment of such astronomical theories on the 'why' of the major extinctions. My purpose here is merely to point to the existence of extinction events as important features of ecospheric history.

Another major and repeated event in the evolution of the ecological theatre is glaciation – or the advent of 'ice ages'. It's common to see references to *the* ice age, but in fact there have been

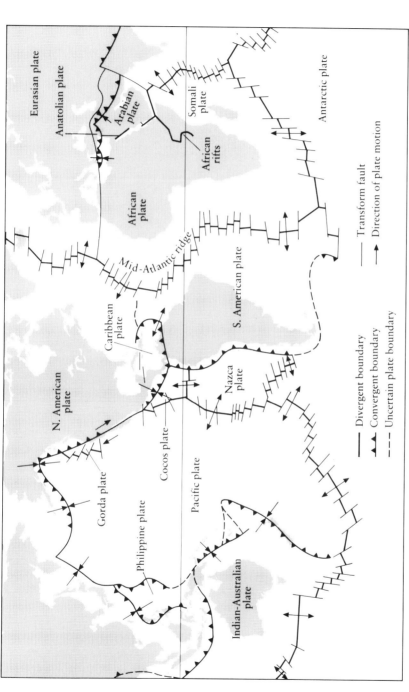

FIGURE 32   Current locations of the earth's major plates.

several. In each case, glaciers spread southwards from the North Pole and/or northward from the South Pole, obliterating or driving before them the flora and fauna of the regions that they cover.

Several connections can now be made. First, glaciations are related to plate tectonics. They appear to be initiated by the movement of a continental area across a pole. When this happens, the land-mass cools and glaciers form. Second, glaciations can be linked with extinction of terrestrial animals and plants for the obvious reason that they destroy their habitats. Third, glaciations can also be linked, though in a less obvious way, with extinction of marine organisms. The connection here operates through variation in the sea level. Because glaciation locks up large bodies of water in the form of ice-masses, a widespread glaciation causes the sea level to fall. While this need not affect deep-sea communities, shallow seas may dry up completely, with consequent extermination of their biota. So for two rather different reasons glaciations would be expected to be agents of extinction.

While there is evidence that some glaciations and mass extinctions coincide[38] (for example in the 440- and 370-million-year-ago events), there are other cases in which extinctions appear not to be accompanied by climatic cooling. And in general we shouldn't expect to find any single cause for the mass extinctions. All have doubtless been associated with some traumatic event in ecospheric history, but glaciations are hardly the only trauma to which the evolving theatre has been exposed. One other that is worth brief mention, and that is in a sense the opposite of glaciation, is the achievement of unusually high temperatures in the centre of Pangaea, in the times when all continents were locked together, and some parts of the land therefore much further removed than usual from the cooling influence of the sea.

The most recent traumatic event in the ecosphere's long history is undoubtedly our own appearance as a major ecological force. From a global viewpoint, it isn't the evolutionary emergence of the hominid line that was crucial – indeed, that could be described as just another 'ordinary' speciation, like the millions that went before it. Rather, what has transformed the ecological scene is the global spread of modern man together with associated destruction of forests, merging of biotas (by accidental and purposive inter-

Realm transport) and modification of the physical environment by various forms of pollution. However, I won't treat these issues here, because to do so would be to get out-of-step. Human and green issues will be dealt with in Chapters 15 and 16. Before we leave evolutionary matters we have a final topic to deal with: evolutionary mechanics. So far, we've only dealt with the 'what' and 'when' of evolution – what major groups appeared at which points in time, for example. But I've managed so far to avoid any mention of the 'how', such as natural selection. It's now time to remedy this omission.

13

OBEYING THE RED QUEEN

If you embarked on a three-year study of a particular ecological community, it's unlikely, to say the least, that both it and you would be enveloped by a mass extinction. The 'big and rare' category of evolutionary events can, from the viewpoint of the practising contemporary ecologist, be disregarded. However, should this fact lead us to put all evolutionary matters to one side as generally educational with regard to ecospheric history but irrelevant to the functioning of present-day communities, we would be making a very big mistake indeed.

One of the clearest conclusions to emerge from the work of ecological geneticists over the last few decades is that evolutionary changes can take place over a much shorter timescale than had previously been thought. In Darwin's day, and for a considerable period after him, a thousand generations was reckoned to be a pretty minimal period for evolutionary changes to occur in. However, while we might still accept such a view in relation to the origin of a new species, it has become clear that significant evolutionary changes can take place on timescales at least one and probably two orders of magnitude less than that – that is, down to as little as ten generations.

Examples supporting this claim are not hard to come by. The evolution of certain moths (and other invertebrates) from populations consisting almost entirely of lightly-pigmented individuals to populations consisting almost entirely of dark ones – in response to the blackening of their background by industrial pollution – took around a hundred years, and so a hundred generations in those species which are 'annuals'. The evolution of resistance to insecticides such as DDT has been observed to occur, in a number of species, on timescales of the order of ten years. And it's now regarded as a plausible hypothesis (though only that) that three-year cycles in the population numbers of some small mammals are partly driven by cyclic evolutionary changes in the populations concerned.

What all this means is that there's no clear line of division between what are sometimes called 'ecological time' and 'evolutionary time'. These labels have been used rather loosely, but roughly speaking 0–1000 years tends to be thought of as the ecological timescale, covering everything from an act of predation (which could be over in a few seconds) to the longest primary succession. That leaves 1000 years and upwards as evolutionary time, in which events like speciation take place. But if populations can evolve significantly in periods measured in decades or less, then it's apparent that restricting our attention to ecological time does not enable us to disregard the possibility that the populations we're dealing with will be very different at the end of the period concerned from what they were at the outset.

In this chapter, I'll be concentrating on evolutionary events operating in ecological time or at the short end of evolutionary time rather than in the very long term; on evolutionary *mechanisms* rather than patterns; and on kinds of evolutionary change that relate most directly to ecological processes such as population regulation, competition and predation. That is, we'll be doing an ecologically-biased sampling of evolutionary biology, as is appropriate. When we've done it, we should be in a much better position to understand features of the natural world that involve both ecological and evolutionary processes, such as the 'latitudinal diversity gradient', which is dealt with in the following chapter.

Most populations of most species are evolving in some way or other most of the time. Admittedly many of the changes may be slight, may be diluted or overridden by change in the opposite direction in subsequent generations, and may be invisible from a morphological viewpoint – but they're occurring nevertheless. Why should such a constant evolutionary flux take place? The answer lies in the impermanence of the environment – both biotic and abiotic – in which all organisms find themselves. If you're perfectly adapted to the prevailing environmental conditions at one moment in time (a rather unlikely situation: see below), then if you don't evolve you almost certainly *won't* be perfectly adapted to the prevailing conditions even a handful of generations later.

The American biologist Leigh Van Valen coined the phrase 'Red Queen Hypothesis' to describe this view of evolution.[39] The idea is that the evolutionary world is like the world of the Red Queen in *Alice Through The Looking Glass*, where you have to run to stand still. Any species that doesn't evolve will find its degree of adaptedness progressively deteriorating, the penalty for which is premature extinction.

The biotic and abiotic components of the environment experienced by any given species operate rather different sorts of Red Queen effect. Essentially, the difference lies in one-way versus two-way causality. If the climate changes, this affects the degree of adaptedness of a species experiencing it. However, any evolutionary response in the species concerned will not have a significant effect on the climate. This is not, of course, to say that organisms don't modify the climate generally – it's quite clear that they do. But rather, a small evolutionary adjustment to two or three enzymes in a species of insect, for example, which makes them more tolerant of high temperatures, is hardly likely in turn to modify the temperature.

When the interaction between a particular species population and its biotic environment is considered, however, causality *can* operate in both directions. A good example is provided by the evolution of predator–prey systems. Here, evolution of avoidance behaviour in the prey may cause selection for increased efficiency of the predator and vice versa.

By describing the totality of ongoing evolutionary changes required to stay equally well adapted in an ever-changing environment as 'obeying the Red Queen', I'm using this term in a broader sense than Van Valen originally intended. However, I think that's no bad thing. Certainly, viewing populations as continually adjusting to altered conditions of life leads us away from the pit of thinking of populations as fixed, genetically inert entities which only undergo evolutionary changes at very infrequent intervals in response to particularly dramatic one-off ecological events.

An alternative mental pit into which the Red Queen *might* lead us is that of thinking that, because of their ability to make continuous evolutionary adjustment to change, populations are always perfectly suited to their environment. This is most unlikely to be the case. Rather, since environments can change more rapidly than evolutionary responses can track them, it's perhaps just as likely that, if a population is perfectly adapted to anything, it's a set of conditions that has recently ceased to exist.

The Mechanics of Natural Selection

The idea of natural selection was brilliant in its simplicity. And in a way it can be attributed to ecologists: Charles Darwin and Alfred Russel Wallace fit better into that category than they do into most others. Certainly they were no geneticists. Since all the kinds of evolutionary change I'll describe in this chapter are adaptive ones driven by natural selection, it's worth a quick tour through the basic idea of how selection operates.

Organisms can conveniently be mentally split in two, into the genotype (the DNA, containing the instructions for building the organism) and the phenotype (everything else). Ecologists tend to be interested in the evolution of phenotypic characters such as body size, life-history patterns and niches, rather than in the evolution of the genetic material itself, which is more the province of geneticists. Of course we assume that a niche or body length or whatever is at least partially heritable: if it wasn't, then natural selection couldn't do much to it. Luckily, what data there is on this matter suggests that at least some of the variation between individuals in such characters is indeed genetically based. Given that fact, we don't have to worry too much, as ecologists, about the precise form that the genetic basis takes.

Most phenotypic characters are continuously variable. That is, there are no discrete categories; rather, individuals merge imperceptibly with each other. So cases like Mendel's famous peas, where there were definite categories – tall and short, for example – are the exception rather than the rule. Virtually all species exhibit continuous variation in size rather than a few size categories, and the same sort of generalization can be made for niches and most other ecologically relevant characters.

Where continuous variation is found, each individual has a *value* and the population as a whole has a *distribution*, usually with a central peak gradually giving way to flanking 'tails', and often doing so symmetrically in the bell-shaped pattern that we call the normal curve. Selection can then be visualized as modification of this curve over one or more generations. Two main types are recognized: directional and stabilizing. In directional, individuals on one side of the mean are favoured at the expense of those on the other side. In stabilizing, individuals close to the centre are favoured at the expense of those in the tails. These processes are illustrated in Figure 33. It's also possible to have selection favouring both tails over the mean ('disruptive selec-

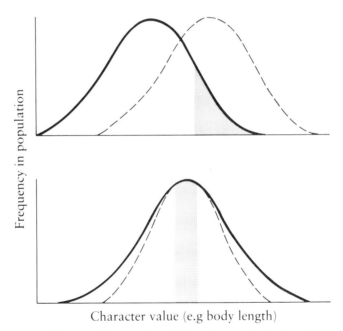

FIGURE 33 Directional (*top*) and stabilizing (*bottom*) selection. Solid curve – prior to selection; shaded area – fittest portion of original population; dashed curve – after selection.

tion'), but that's thought to be comparatively rare.

Stabilizing selection tends to keep the population where it is. A frequently cited example concerns birth weight in humans. On average, very light and very heavy babies have slightly higher mortality rates than those close to the average. Consequently, the average tends to stay put. Directional selection, on the other hand, pushes the population's character distribution towards generally higher or lower values, depending on which are favoured. An example is provided by Bergmann's Rule, which says that in many species of mammals, the individuals get bulkier as you move towards higher latitudes. Presumably, the basis of this is that in cold climates selection favours bigger individuals, with lower surface area/volume ratios, because they lose their body heat to the environment less quickly. If this sort of process were to operate on a species of mammal spread from northern Scandinavia to southern Spain, it would produce a latitudinal 'cline' in body size from initially similar-sized colonists.

Because environmental conditions are different in different places, widely distributed species are often broken up into geographical races by selection favouring different combinations of characteristics in different areas. The human races are an obvious example, even though in them, as in many other species, it's difficult to see the selective advantage of all of the characters that vary, and there's a strong suspicion that some of them are only changing because they're genetically or developmentally linked to others that are ecologically more important.

Occasionally, we find phenotypic characters that vary in a discrete way – for example, blue versus brown eyes, banded versus unbanded snail shells, black versus pale moths, and so on. The reason for the discrete nature of the variation in these cases is that instead of the genetic basis of the character concerned being lots of genes each with tiny effects that blur into one another, it is instead just one or a few genes with relatively major effects. In such cases, selection can again be directional or balancing, but the way we picture it is different. Rather than there being a continuous distribution of character values, the population is characterized by particular percentages of the different phenotypes or 'morphs'. Balancing selection is then thought of as keeping the population polymorphic, while directional selection, by favouring one morph at the expense of others, drives the population towards being composed of 100 per cent of the fittest phenotype.

I should stress that all of the forms of selection that I've mentioned so far are forms of *individual* selection. That is, the process operates through differential fitnesses of genetically-different individuals. For example, a population's mean body size may increase over several generations because, in the prevailing environmental conditions, large individuals on average leave more surviving offspring than small ones. However, it is possible to envisage selection as operating at different levels than that of the individual. In particular, some evolutionary processes might be driven by group selection, wherein fitness differentials occur between groups, or populations, rather than between individuals within a single population. I'll postpone discussion of this, as it links in naturally with the evolution of 'self-regulation', which is dealt with in a later section.

As I hope I've made clear, the idea that natural selection can readily turn a species into a patchwork of genetically distinct populations and races is a simple and unobjectionable one. What's not nearly so simple, however, is how such races in some cases eventually break reproductive contact and become separate species in their own right. Yet this process of splitting or speciation is crucial, as it produces the very entities whose interactions the ecologist is interested in. So we need in particular to examine the origin of species – that subject that, ironically, Darwin came close to omitting. But since speciation is difficult, let's leave it to last.

Tracking Environmental Change

We'll deal first with the simplest kind of evolutionary change – that in which the population is responding to a change in its physical environment. A good example of this is provided by the evolution of heavy metal tolerance in plants. Mining of heavy metals such as copper, lead and zinc results in the creation of contaminated areas in the vicinity of the mine where metal-rich 'spoil' is dumped. Often, the substrate of these contaminated areas is so toxic that nothing will grow on it. However, since some of them have now been around for more than a hundred years, there has been time for some species to evolve metal-resistant strains that can colonize these inhospitable environments.

One celebrated case-study involves the evolution of lead tolerance in the grass *Agrostis tenuis*, studied in detail by the British ecologist Tony Bradshaw and his colleagues, in the vicinity of an old lead mine in Wales. Here, testing of lead tolerance in plants taken from various points along a transect revealed a dramatic increase in lead tolerance on entering the contaminated area. Presumably, genetic variation in lead tolerance in the original population had been acted upon by directional selection as indicated in Figure 33, gradually producing a more lead-tolerant strain. An interesting aside here is that lead-resistant plants flower earlier than lead-susceptible ones, so we can see a possible mechanism for the breakdown of reproductive contact in the future, though I should stress that this hasn't yet happened, and the tolerant variety is thus still firmly within the parent species.

This link between a physiological and a reproductive character raises the general issue of 'correlation of characters', as Darwin called it. For various reasons, including overlap in their genetic bases, different phenotypic characters often exhibit joint responses to selection: if you modify one, you automatically modify the other too. There has been much criticism of the tendency, on the part of ecological geneticists and others, to 'atomize' organisms into a number of phenotypic traits, and to treat the evolution of each of these in isolation from the others. Clearly, such an approach is overly simplistic, but at the present limited stage in our knowledge of character correlation it's often the only approach possible. In the future, as our knowledge improves, we should be able to get to grips much more confidently with 'multi-character evolution'.

Occasionally, there isn't the right kind of variation pre-existing in a population for it to respond to a sudden environmental change. One possible result of such a state of affairs is extinction. However, another is that a new mutation will occur, providing the kind of variation upon which selection can act. Novel mutation may have been a contributory factor in the evolution of lead tolerance in *Agrostis*, though we can't be sure about that.

Directional selection in response to altered environmental conditions is a very widespread process in natural populations. Sometimes it occurs because the population stays put but the environment it's in changes. In other cases it occurs because the population migrates into areas where conditions are already different. This second possibility must frequently be manifested as a species expands its range from small beginnings. Human races, and their equivalents in many other geographically widespread species, are caused in just that way.

Tracking the Opposition

Things become slightly more complicated when a population, instead of responding to something such as the substrate or climate that is oblivious to its presence, evolves in response to something that can 'evolve back' like a competitor or predator. This kind of situation, where ecologically interacting species exhibit interacting evolutionary responses to each other, is referred to as coevolution.

Predator–prey or herbivore–plant coevolution has been called an evolutionary arms race by the well-known British behavioural ecologist Richard Dawkins, whose recent book *The Blind Watchmaker*[40] deals thoroughly with this topic. The sort of process that Dawkins envisages as an arms race can be exemplified by the coevolution of a plant and a herbivorous insect which consumes its leaf tissue. Plants which are not heavily grazed upon will leave more offspring to future generations than those that are. Consequently, any genetic variants within the plant population that are less palatable to insects will increase in frequency. It is in this way that many plants have come to contain all sorts of toxins, even including (for example in some clovers) cyanide. However, as a plant population becomes predominantly composed of plants containing toxic substances, this creates a selection pressure on the insect population favouring those variants whose biochemistry is such that they can still consume the plant tissue despite the toxin –

e.g. by having an enzyme system that converts it to something harmless.

'Arms race' is perhaps a misleading label for some types of predator-prey coevolution. For example, if we turn from insect–plant to lion–zebra coevolution, an obvious feature of the system is selection for faster prey causing selection for faster predators, causing yet more selection for even faster prey, and so on. This is hardly an arms race, since the prey's speed is a means of escape, not of attack. This contrasts with the plant's toxins, which definitely do attack any insects which attempt to consume them. Nevertheless, whatever label we choose to put to any particular case of predator–prey coevolution, there is a common theme running through these coevolutionary systems, namely the mutual rather than one-way causality to which I referred earlier.

Of course, not all coevolution occurs in the context of predator–prey systems. Another context in which mutually-induced evolutionary change is to be expected is interspecific competition. However, competitive coevolution has proved to be a rather elusive process, as those who have attempted to study it (including myself) know all too well. If two species compete for a limiting set of resources, two quite different kinds of evolutionary response can be envisaged. The first, called character displacement, is where individuals in either species that are least like the other in their ecological requirements suffer least competition, and are consequently fittest. This would lead to selection 'pushing apart' the niches of the two species, as illustrated in Figure 34, and also the mean values of any other phenotypic characters that are associated with niches. Such a process, it should be noted, assumes that some of the variation in a population's resource use that a niche represents is due to genetic differences between individuals, over and above any variation that is due to behavioural flexibility. That is, the population's niche is broader than the individual's.

There are lots of proposed examples of character displacement, but few if any are conclusive, so it remains to be seen whether or not this coevolutionary process is widespread in natural communities. If it is, then it opens up a way of achieving regular dispersion of niches along resource axes other than by competitive exclusion, because if species were initially clumped, a bit of 'coevolutionary

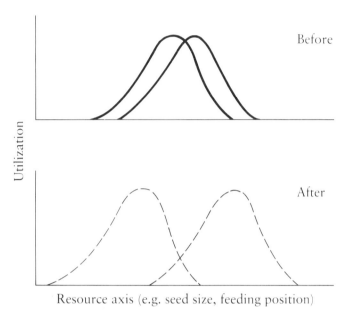

FIGURE 34   The coevolutionary process of character displacement, leading to an increased niche difference between a pair of competing species.

shuffling' could have a dispersing effect. If it isn't, then we have to ask what alternative kinds of evolution competition is likely to give rise to, since it seems inconceivable that it causes none at all. The most obvious alternative to shifting your resource base is becoming better at utilizing your existing one. For example, selection may favour individuals that are more efficient at finding resource items. This kind of change, occurring in either or both competing species, is generally referred to as evolution of increased competitive ability. Interestingly, while character displacement increases ecological stability by decreasing the threat to it posed by competitive exclusion, evolution of increased competitive ability is potentially *de*-stabilizing. So one important question for future research is which kind of competitive coevolution predominates in nature. For what it's worth, my guess is that it's the de-stabilizing kind.

I shouldn't give the impression that all coevolution is based on conflict. Predator–prey arms races and competitively-induced changes certainly are conflict-based, but there's a lot of coevolution associated with mutualistic interactions too. Coevolution of flowering plants and their insect pollinators is one of the commonest examples of this, but *all* cases of mutualism, from root-nodule bacteria to ant-harbouring acacias, involve quite considerable coevolutionary adjustment on the part of the species concerned.

The Evolution of Self-Regulation

I said in Chapter 9 that, subject to a qualification that I'd take up later, resources and enemies were the only ecological factors that could act as sources of density dependence, and hence of population regulation. What's the qualification? Well, essentially, it's that these are the only 'extrinsic' ecological factors that can regulate. But a population is also capable of regulating *itself*. So as well as the possibilities of resource-regulation and predator-regulation, we have to consider the additional possibility of self-regulation. The idea of self-regulation will only begin to make sense if I give some examples. In other words, we have to examine, in particular cases, *how* a population can self-regulate. One of the best examples is territoriality. This used to be thought to be typical only of certain bird species, but it is now known to occur also in other vertebrates, and indeed in some invertebrates too – such as the crabs which defend sectors of beach as 'their own'. Territoriality is actually a very widespread process, both taxonomically and geographically.

There is little doubt that a population can be self-regulated through territorial behaviour.[41] Let's consider how this could happen in a bird species – such as the famous red grouse – inhabiting an area of moorland, as illustrated in Figure 35. Given a fixed overall area of suitable habitat, and a given territory size that each male will defend (though there's some variation in this), the total habitat area becomes a sort of jigsaw of non-overlapping but abutting territories. The maximum population size of territory-holding males is therefore set at whatever number you get by dividing habitat area by territory area. Males that are ousted from

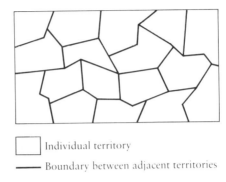

Individual territory

──── Boundary between adjacent territories

FIGURE 35 'Jigsaw' of grouse territories on an area of moorland.

the region because of losing territorial battles either don't breed or migrate to other regions, in either of which cases they cease to be effective members of the local population with which we're concerned.

Clearly, any temporary excess of males will subside, while deficiencies, in the form of suitable but unutilized space for territories, will soon fill up with immigrants or maturing juveniles. So we have a thermostatic or density-dependent system, which will produce an equilibrium in much the same way (in terms of system behaviour) as one in which the density dependence stems from food or enemies.

Let's now turn from the phenomenon itself to its evolutionary origin. Resource-limited populations can sometimes over-exploit their food supply and 'crash' – that is, they go beyond the realm within which density dependence is able to stabilize them, and the result, in extreme cases, is a local extinction. Territorially self-regulated populations will be nearly immune to this sort of problem. So we can see an obvious advantage to a population in being territorial, and so an advantage to the species generally.

Now on a naïve view of how natural selection operates, we could be tempted to argue as follows. Territoriality is advantageous to populations because it reduces their probability of 'crashing'. Therefore, if in an initially non-territorial population a genetic novelty turns up through mutation or recombination that permits the individuals carrying it to be territorial, the population will evolve towards an all-territorial constitution, because this will be to its benefit.

What's wrong with this kind of reasoning is that it involves the spread of a new variant because of its potential benefit to the population at some time in the future. But natural selection doesn't work in that way. Rather, it works through differential fitnesses between individuals within a population which affect what happens in the present. Arguing that the potential benefits of self-regulation explain why territoriality arose is in essence arguing that a cause can come after its effect.

The invalid argument we've just disposed of is labelled 'group selection' – though we have to be careful here because that phrase is used in more than one way. The true explanation of the origin of territoriality (as far as we know) is that when the first few territorial variants arose in a largely non-territorial population, those individuals held territories, and consequently prospered more, in terms of surviving offspring, than other individuals, because of the steady supply of resources that a territory ensures. So each generation the frequency of the territorial type increased a little until eventually it displaced the original type entirely. At the end of the process, it's quite true that the population as a whole may have a better future because of its ability to self-regulate, but that's an accidental side-effect of the evolutionary process, not its cause.

There is, however, a kind of group selection which may come into play later on. If we suppose that suitable genetic novelties for territorial behaviour occur in some of a species' populations but not others, the species as a whole will become a patchwork of local populations some of which are territorial and self-regulating, others not. Given that the former will on average be more enduring, a long period of sporadic local extinctions here and there followed by recolonization of the areas concerned by migrants from other populations would be expected to increase the frequency of self-regulating populations within the species until eventually all were of that kind. But this second kind of group selection, while more plausible than the first, can only operate *after* the initial spread of the appropriate genetic novelty through at least one population on the basis of its benefits to individuals and their families.

The Origin of Species

Speciation, like self-regulation, is a very important process, yet one which is really an accidental spin-off from something else. It's often difficult for newcomers to the field of evolutionary ecology to get to grips with the fact that, because of the way natural selection works, many of the most important things it achieves were not, in a sense, what it was 'trying to do' at the time. In the case of speciation, the problem is even greater than before, because we're dealing now with one of the most important evolutionary events of all. Everything that ecologists do involves species. Whether we're looking at population growth, competition, predation, community structure or succession, we're compiling data on species. Yet the key process that brings these all-pervasive entities into being is an entirely accidental one.

While there are several ways in which speciation can come about, the one that's thought to be commonest (at least in the animal kingdom) involves the evolutionary divergence and eventual reproductive separation of a 'peripheral isolate'. This phrase refers to a small outlying population which has come to occupy, through an unusually dramatic, one-off migrational event, a geographical location far from the main range of its parent species.

Since speciation has turned out to be an elusive process, with the result that all our case-studies of it happening are somewhat fragmentary, I'm going to generate a speciation event in our fictitious blue-tailed thumper in order to illustrate how we think the speciation process works. I hope that the benefit gained in terms of formulating a complete picture of speciation will outweigh the scientific dubiousness of giving precedence to fantasy over fact.

We'll start by supposing that our single species of blue-tailed thumper initially consists of a patchwork of populations inhabiting the Kerulen Valley system – that is, the valley of the Kerulen river, and the 'side-branch' valleys of its tributaries. As in any species, there's considerable genetic variation within and between the various populations, but nothing yet as dramatic as a pronounced geographical race.

Now suppose that a small group of thumpers are grazing on lichens on the trunk of a fallen tree that's floating adjacent to the bank of the river. A sudden current moves the log away from the bank, and it begins to float downstream. Thumpers being poor swimmers, they decide to remain 'on board'. The log is carried out to sea, where it drifts southwards for several days, and is eventually washed up on the beach of a small Pacific island. The whole scenario sounds wildly implausible, of course, but no more so than many real migrational events: there are factual accounts of it 'raining snails' in tropical storms.[42]

If our few migrating thumpers survive their arduous journey, they'll no doubt clamber eagerly off their unwanted vehicle and begin to explore their new environment. If it's hospitable – i.e. provides sufficient food and not too much in the way of enemies or other threats – they'll prosper and, assuming the migrating band contained both males and females, breed. The population will grow and eventually stabilize like any other.

Concurrently with the ecological process of population growth, an evolutionary process of selective adjustment to their new conditions will begin to operate. The peripheral isolate population (as it now is) will be genetically variable though, having said that, it may be less variable than its parent population, because of having gone through a 'bottleneck' of low numbers. It may also contain different relative frequencies of particular variants to the population from which it arose, simply because the migrating individuals comprise a small sample, and, as we know, small samples can be quite dramatically non-representative.

Given, though, that there's *some* genetic variation, natural selection will begin to increase the frequencies of those variants best suited to the new conditions. The prevailing temperature of the new environment is higher and the array of resources available quite different, so all sorts of phenotypic characters, from 'invisible' enzyme systems to the body size, the thickness of the fur and the feeding niche, will begin to alter. After many generations, directional selection will give way to stabilizing, as the population's genetic structure reaches as close to optimal for the new environment as it can – though small changes will continue in response to temporally-varying conditions.

As we can see, the driving force here is the 'need' for local adaptation of the population, and this is achieved by straightforward directional selection. The end result of such a process *might* just be a well-marked geographical race of the parent species, but things get more interesting if – and we'll assume that this is the case – the genetic bases of ecologically important characters (such as those mentioned above) overlap with the genetic bases of reproductive characters. If this kind of character correlation exists, then the selection that modifies body size, for example, may unwittingly also modify such phenotypic traits as courtship behaviour, which will of course affect reproductive compatibility of the peripheral isolate thumpers with parent-population thumpers, should the two groups ever re-contact and have the opportunity to interbreed.

Suppose that there are two main reproductive spin-offs. First, male courtship behaviour changes so that the island males give double thumps instead of single ones to attract their mates. Second, chromosomes are altered so that hybrids between island and mainland thumpers (none of which have occurred yet) have a high incidence of sterility – though not total sterility as in interspecific hybrids like the mule (a horse/donkey hybrid).

Now if the mainland population expands its range southwards and eastwards so that it comes to occupy those areas of coastal mainland nearest to the peripheral isolate, a second migrational event might well bring a propagule of island thumpers back into contact with mainland ones. What will happen then? Inter-type mating will take place, but because of some hybrid sterility there can be said to be partial reproductive isolation. Then the so-called Wallace Effect (after Alfred Russel Wallace) will begin to operate, as follows. Those females in either group which are more attracted to 'own-type' males will leave more fertile offspring than those which prefer 'other-type' males, and their genes will increase in frequency. The culmination of this process is the stage where complete reproductive isolation is achieved because each group consists only of individuals which won't mate with the alternative group. At this stage, speciation is complete, and the two species as they now are will go their own ways, both ecologically and evolutionarily, even if they inhabit the same geographical regions.

Coming back to reality, it's easy to find examples of all stages in the process, but in *different* species. Countless species have geographical distributions restricted to single regions, with no peripheral isolates, while many others do have small outlying populations in far-flung places. Occasionally, for example in the fruitfly species *Drosophila pseudoobscura*, we find, by attempting crosses in the laboratory, that a peripheral isolate has become partially reproductively isolated from its parent form, thus illustrating the beginnings of speciation. And the existence of pairs of sibling species that are still capable of very occasional interspecific mating – like the snails *Cepaea nemoralis* and *C.hortensis* – shows us what are essentially speciational end-products. But nowhere is there to be found a real case-study showing us all stages of the process: hence the resort to thumpers.

Having seen at least one of the ways in which species can come into being, we're now a little better equipped to deal with ecological patterns of species richness such as the 'latitudinal diversity gradient', which is our next port of call. One possibility is that variations in species diversity from place to place are caused by variations in the speciation rate. But is this the correct explanation, and if not what is? The next chapter won't give a cut-and-dried answer to this question, because ecologists don't yet have one, but it should, with luck, point us in the right direction.

14

Nature's Most Impressive Pattern

Let's go for a 10,000-mile walk. What I have in mind is a 'walk of the Americas', starting in the Northwest Territories of Canada somewhere in the vicinity of the Arctic Circle, and taking a meandering path southwards, ending in the Amazonian rainforest a stone's throw from the Equator. Unlike some previous long-distance hikes, the purpose of this one is not to demonstrate our fitness or powers of endurance. Rather, its purpose is to allow us to observe, at intervals along our chosen route, the resident biota of the areas concerned, and in particular to estimate the number of species that we find there – that is, the species richness or diversity.

Of course, in reality this would be a fiendishly difficult task. As anyone who has tried to identify an insect collected from a pond knows all too well, it's often almost impossible to allocate a specimen with confidence to a particular species category. Imagine the problems involved in carrying out this exercise for all the specimens we collect on our walk. First, there's the enormous scale of the task, involving thousands of species. Second, there's the fact that the species concerned represent lots of different taxonomic groups, scattered throughout the plant, animal and microbial worlds. Finally, there's the problem that many of the areas through which we pass – especially towards the end of our journey – will be very poorly known and may well contain species which have yet to be discovered and described.

It seems prudent to take whatever steps we can to decrease these difficulties. Let's take along a posse of tame taxonomists (if there are any left), made up of specialists in all of the groups of organisms that we're likely to encounter. Also, let's simplify our sampling programme by taking a route that sticks as far as possible to lowland areas. This will help to avoid the problem of confounding altitudinal and latitudinal trends. And for the moment, let's focus our attention primarily on forest communities, wherever we encounter them, keeping data on other community-types (grassland, scrub, lake, and so on) for subsequent analysis.

Suppose, then, that we accumulate information on species identities in a series of roughly equal-sized and equally spaced sampling areas, each separated from its neighbours by about 500 kilometres, and each measuring 10 kilometres square. If we sit down at the end of our journey, tot up the total richnesses for each area and compare them, what sort of pattern, if any, shall we find?

I suspect that most people, even those with no ecological training whatever, will have an inkling of the answer, but not, perhaps, a grasp of the magnitude and all-pervasiveness of the trend that is found. Our major finding will undoubtedly be that, despite whatever 'noise' there is in the data, the number of species increased as our journey progressed, being lowest at our northern starting point and highest at our Amazonian ending. Moreover, this pattern of increasing diversity with decreasing latitude – the latitudinal diversity gradient – will be found in all major taxonomic groups. It isn't something that's restricted to plants or birds or insects (for example): it will be found in all of them.

Of course, we have to be careful to include, as I have done, the qualification of *major* groups. If we restrict our attention to the 'smaller' taxonomic categories, like families or genera, then we can easily find groups that show a trend in the opposite direction to the overall pattern, because of the specific type of adaptations exhibited by the groups concerned. If I claimed that the species diversity of penguins increased on moving from pole to equator, for example, you might find that a little difficult to swallow. But birds as a whole, including penguins, do indeed show a marked latitudinal trend in the 'right' direction.

Not only does the latitudinal diversity gradient prevail in all major taxonomic groups, it also prevails in all major habitats. I said that to begin with we'd restrict our attention to forests, and these certainly exhibit one of the clearest trends, with an area of boreal forest being vastly less rich in species than a rainforest area of similar size. But the gradient is by no means restricted to forests. Grasslands show a similar trend, as do other terrestrial communities; and most aquatic ones behave likewise, though there's some uncertainty in the case of a few types of freshwater system.

As well as being taxonomically and ecologically pervasive , the gradient is geographically widespread too. Instead of a 'walk of the Americas' we could equally have trekked from Norway to the Congo or from eastern Siberia to Borneo, and the result would have been the same. Alternatively, we might have opted for a Southern Hemisphere hike. This presents some problems for the student of terrestrial communities – though it's great for marine data – because it isn't possible to take a continuous land-based route from the Antarctic Circle to the Equator, and there will consequently be latitudinal gaps in our overall picture. Nevertheless, the gradient would still be found.

There are two reasons why I've chosen to call the latitudinal diversity gradient 'nature's most impressive pattern'. The first is its grand geographic scale and its all-pervasiveness. The second is simply the magnitude of the changes involved. For example, if we compare equal areas of boreal forest and tropical rainforest (a hundred or so square kilometres apiece), we might find ten species of tree in the former and perhaps a thousand in the latter. Often, ecological patterns are relatively weak, and we need to resort to statistical techniques to test whether the patterns are really there or whether we're imagining them. But the latitudinal gradient is so dramatic that statistical testing is irrelevant. Let's have a look at some numerical examples in selected taxonomic groups, to see just how impressive the pattern can be.

Same Pattern, Different Beasts

We'll start with birds, whose species diversity is a good deal easier to estimate than that of most other kinds of animal. Figure 36, taken from Robert MacArthur's fascinating book *Geographical Ecology*,[43] shows the number of breeding species of land birds in North and Central America. The number varies from about 20,

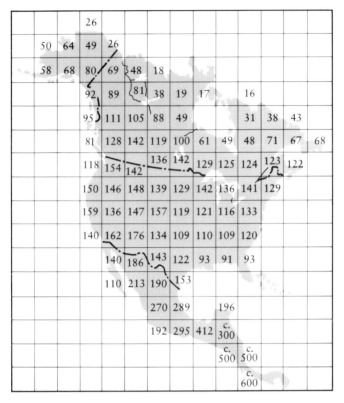

FIGURE 36   Number of species of breeding land birds in North and Central
America.

along the northern coastline of Alaska and Canada, to about 600
in Costa Rica, so there's about a 30-fold difference.

Similar sorts of pattern are illustrated in Figure 37 for two very
different kinds of animal: lizards and 'calanoid copepods'. Lizards
are of course exclusively terrestrial, and the data-set shown derives
from North American studies, which makes it broadly comparable, geographically, to MacArthur's data on birds. Calanoid
copepods, in contrast, are a group of marine planktonic crustaceans, and the data-set illustrated relates to a latitudinal 'transect'
from tropical regions of the Pacific right up to the Arctic Ocean.
The lizards show about a 15-fold increase in diversity over the

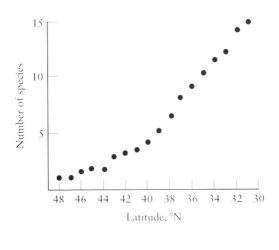

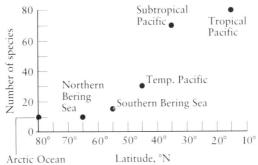

FIGURE 37   Latitudinal gradients in species richness. *Top*: lizards in North America. *Bottom*: calanoid copepods (a kind of crustacean) in the Pacific and Arctic Oceans.

relatively limited latitudinal range concerned, while the copepods show a less marked 7-fold increase across a broader range. But in all cases, there can be little doubt that the trends are real; and they are clearly all in the same direction – more species as we move towards tropical regions.

Information is available for many more groups than those I've illustrated here, but little is to be gained by inspecting more and more examples. Rather, having seen the general form that 'nature's most impressive pattern' takes, we should turn our attention to seeking an explanation of *why* this pattern occurs, rather than any alternative one. We'll move on to the difficult business of considering competing explanations in the following section, but first I should briefly mention why the comparisons I've been making have always involved equal areas.

If we restrict our attention to one particular kind of organism and to one particular place, but take an increasingly large area of ground into consideration around some central study-area 'marker', then up to a point we find an increasingly large number of species. For example, suppose we're conducting an investigation into the number of species of land snail in Great Britain, and we start somewhere in the middle, at a randomly chosen spot in rural Derbyshire. Taking a sampling radius of 1 centimetre around a post we hammer into the ground to mark our spot, the species count will probably be zero. At 1 metre, it may well be one or two. At 100 metres, we'd expect to find several species, though exactly how many there would be would depend on what kind of terrain our initial reference point turned out to be in. At 10 kilometres, the species count would be higher again, with between-habitat differences in the molluscan fauna beginning to account for part of the rise.

The tendency of species richness to increase with area is often referred to as the species-area curve, and much work has been invested, by ecologists, in characterizing this curve in detail for particular groups in particular places. Its relevance here, of course, is that it provides the rationale behind taking care to compare equal-sized areas when claiming evidence of species-diversity gradients. Clearly, if a comparison was made involving unequal areas, almost any pattern could be made to appear. But since we know that, we can ensure that it doesn't happen.

One final quibble: people who don't like nice neat patterns are prone to argue that deserts constitute an exception to this one. While it's true that deserts represent fairly low-diversity systems and are mostly located in tropical or at least low-latitude areas, that's no reason to propose that they 'disobey' the general latitudinal gradient. The appropriate comparison is of the same type of habitat at different latitudes. Hence our consideration of boreal versus temperate versus tropical forests or arctic versus mid-latitude versus tropical oceans. In the case of deserts, no very broad-scale comparisons can be made because there aren't any deserts in arctic regions. But deserts *do* occur at different latitudes (e.g. South Sahara at about 20°N, compared with western North American deserts at 40°N) and, although I haven't seen any comparative data on deserts at different latitudes, I suspect that such information is available somewhere, and that it shows a latitudinal trend.

From What to Why

In my experience, ecologists go through three phases during their 'professional development', in their attempts to seek an explanation for the latitudinal diversity gradient. The first phase embodies an attempt to come up with a single, simple cause. One such causal agent often proposed is the undoubted latitudinal variation in the amount of incident solar energy. The argument, if it can be called that, goes something like this: tropical regions receive the most solar energy, therefore they will be able to support more species.

Unfortunately, this proposition does not stand up to analysis. It's true that tropical regions receive most energy, and that will undoubtedly give them the potential for a greater rate of organic production. That is, the rate of laying down of new organic material per unit time will be higher there. Given that there is a general limit to the efficiency of ecological systems, as we saw in Chapter 4, then there *must* be a higher production rate, on average, in tropical systems. The only way this could fail to happen is if tropical communities were systematically less efficient than high-latitude ones, and I can't think of any reason why this should be so. Therefore, applying a similar average efficiency to different rates of energy input will lead to different production rates, with tropical regions highest and arctic ones lowest.

The problem for the 'simplistic' theory of an energetically-induced latitudinal gradient lies in the difficulty of converting production rates to numbers of species. There are two stages (at least) in this conversion, and neither of them work. The first thing to do is to convert temporal rates to instantaneous measures, which takes us from production to biomass. As we saw in Chapter 4, the relationship between production and biomass is complex. It's possible to have a very high biomass being rapidly created and destroyed (high production and consumption rates), or alternatively the same very high biomass with an exceptionally low turnover. In fact, we can have any combination of high or low biomass with high or low production rates.

As if that weren't bad enough, the conversion of biomass to number of species is, despite their both being instantaneous measures, equally problematic. Again, all combinations are possible: that is, high/high, high/low, low/high and low/low. The boreal forest, for example, has high plant biomass but low plant diversity, while a savannah region has the opposite combination.

Given that it depends on not one but two simple relationships, neither of which exists in reality, the explanation of varying species diversity in terms of varying energy input doesn't have much going for it; and the realization of this takes developing ecologists on to 'phase two' in their search for an explanation of the latitudinal gradient. This is the phase of sophisticated but purely ecological theories devoid of any serious attempt to get to grips with the evolutionary dimension of the problem.

It's not too hard to find theories which fit into this category: many ecological texts are full of them. Some of them don't appear to make sense, while others, to some extent, do. Let's look at one example of the 'sensible' category. I'll call this the 'competition theory', though it really involves a mixture of competitive, and other, influences. The basic idea is this. In a more predictable, less variable environment, it is possible to be a narrower specialist, in terms of feeding, than when conditions fluctuate. Also, in less variable conditions, competitors can be more tightly 'packed' together,[44] as we saw in Chapter 8. So both factors influencing the number of species that can be accommodated by a given array of resources, namely niche breadth and niche overlap, will tend to be more conducive to tighter packing in the tropics.

At one level, this hypothesis seems to make sense. It's true that it is based on the idea of guilds being resource-regulated and competitive, which may not be the case in some systems, but where guilds *are* competitive, then it seems a plausible idea.

However, this is where we come up against the problem of the lack of recognition of an evolutionary dimension.

Let me try and get across exactly what the problem is. If it were the case that at all latitudes the numbers of species present in the regional pools were identical, but local communities contained more species at lower latitudes than at higher ones, then it would seem that an exclusively ecological theory – such as the 'competition theory' – would be appropriate. But it's clear that this is *not* the case. Rather, the regional pools get smaller as we head north or south from the equator. Some examples make this particularly obvious: we don't find thousands of species of tree available for colonizing boreal forest latitudes, with only a few of them being represented in any one local patch. Rather, there are far fewer trees available as potential colonists generally at this latitude than in more tropical ones. It would appear that the *main* difference is between regional pools, not between fractions of regional pools that are actually found at particular localities.

Given this fact, we shouldn't look to hypotheses based entirely in contemporary ecological processes to explain the origin of the latitudinal diversity gradient. Rather, we should be thinking in terms of how it is that regional pools in the tropics come to have far more species generally than their temperate-zone equivalents, which in turn are more diverse than those in the Arctic. This conclusion inevitably takes us into the evolutionary realm of rates of speciation and extinction, and the question of why such rates should vary on a latitudinal basis.

Speciation versus Extinction

In any taxonomic group – molluscs, birds, plants or whatever – the number of species can only be altered by speciation and extinction. If we consider the slightly more complex question of what could alter the number of species in one of these groups within a particular band of latitudes, then it's clear that north–south geographical movement of species could also be a contributory factor. (All these, incidentally, have parallels with population dynamics: speciation is 'birth', extinction is 'death', and geographical movement of whole species distributions is 'migration'.)

While we have to acknowledge the theoretical possibility that species-level 'migration' might need to be taken into account, it probably isn't important in practice. Movement in the east–west dimension is reasonably easy, as the existence of many Holarctic species (i.e. those whose distributions extend through both Nearctic and Palaearctic Realms) shows us quite clearly. But there are almost no species with equally wide distributions in the north–south dimension, such as 'Holamerican' ones. The reason for this is almost certainly that because conditions differ so much latitudinally, North-South movement on a broad scale usually causes speciation to happen. Thus within-species large-scale latitudinal 'migration' can be relegated to a very minor role.

What this means is that the origin of latitudinal variation in numbers of species must be sought in latitudinal variation in the number of speciation events or extinction events. Which of these two is more likely to be important? It isn't possible to answer this question with certainty, but I think that speciation is probably the main factor. If it were extinction, then we'd be saying (for example) that there were thousands of species of coniferous tree produced by speciation, but their extinction rate was so high that only a few survived. If this sort of explanation applied generally, we might expect to find evidence of the massive excess of high-latitude extinctions that it would imply in the fossil record. Since such evidence is lacking, we turn to speciation as the more probable cause of the latitudinal gradient.

There are two ways in which high-latitude locations could be characterized by fewer speciation events. Either the speciation rate per unit time is lower or the amount of time available for speciation is shorter, or both. Because of the continental movements and glaciations that we saw in Chapter 12, it does seem that tropical conditions have had a longer unbroken history than temperate and arctic ones. So it's likely that the time factor has had a contributory effect. However, we can ask ourselves the question: given unlimited persistence of present geographical, atmospheric and geological conditions, would the latitudinal gradient eventually disappear and species richnesses right across the globe all come to be as high as they currently are in the tropics?

Since there's no stopping plate tectonics, we'll never be able to answer this question with certainty. If the answer is 'yes', then clearly the latitudinal gradient was a purely transient phenomenon whose explanation lay in one-off historical events. If the answer is 'no' (and I suspect that it is), then the gradient is caused, at least partially, by higher speciation *rates* at lower latitudes.

What would cause such latitudinal variation in the rate of speciation per unit time? This is a difficult question, not because we cannot envisage an answer, but rather because there are too many possible answers, since so many things vary on a latitudinal basis. Which of them, if any, is the main cause of a negative relationship between speciation rate and latitude is, at this stage in our knowledge, entirely a matter of guesswork.

Let's do some guessing, then, however scientifically disreputable that may be. A sensible starting point is to admit that we can only think about what might cause the *rate* of speciation to alter if we have some idea about how speciation usually happens – i.e. its *mechanism*. We saw in the previous chapter that, at least in the animal kingdom and perhaps in the others too, allopatric speciation via peripheral isolates is thought to be commonest. While even this proposal is guesswork, let's assume, for the purposes of the present discussion, that it's broadly correct.

Probably the biggest stimulant to allopatric speciation is geographical fragmentation of species distributions, and restricted movement between the 'fragments'. Might tropical distributions be more fragmented or patchy than others? It's possible to argue that they should be, as follows. Most species have a 'main range' that is inhabited by the species concerned wherever within it suitable resources are found. But always, small peripheral populations will be 'thrown off' by migration, beyond the main range. In a harsh and highly variable overall environment such as the Arctic, local extinction may be the usual fate of such small populations, thus nipping in the bud any possibility of a trend towards speciation. In the tropics, where for the most part environmental conditions are both more generally favourable to life and less variable, peripheral populations may more often survive long enough to locally adapt, and ultimately speciate. You could give this possible mechanism for higher speciation rates at lower

NATURE'S MOST IMPRESSIVE PATTERN

latitudes some grandiose title, like the Spatial Mosaic Theory, but I wouldn't advise it because, like all other current 'theories' of the latitudinal diversity gradient (some of which *have* been given grandiose titles), it's really just a guess, as I said at the outset.

You may find it interesting to try to invent your own theory. If you do have a go at this, one thing worth considering is that whatever mechanism you come up with has to work in all types of habitat. It's easy to forget that what we have to explain is not just the existence of a diversity gradient in forests, but one which is all-pervasive and found, among other places, in the sea. So theories based on soil or structural heterogeneity of the habitat can at best only provide a partial answer to our problem: they cannot claim to solve it on an ecosphere-wide basis.

Beyond the Origin

I've attempted to explain the origin of the latitudinal diversity gradient in terms of patterns in the origin of species. In doing so, I've been somewhat dismissive so far about purely ecological theories of the gradient, like more equable tropical conditions promoting narrower niches and tighter species packing. However, in the end we must come back to such theories, because if, at any latitude, there is a long period of increasing species richness due to the speciation rate exceeding the extinction rate, the newly-arisen species will only persist if there is 'room' for them, in terms of niche space, in the communities of the day. In other words, while a gradient in species diversity is *originated* through a gradient in speciation or extinction rates, it is *maintained* through a gradient in community dynamics. As we've seen already, it isn't yet clear whether the latitudinal gradient *is* maintained or whether it's a purely transient historical phenomenon, though I think the former is more likely.

One thing that discussion of the latitudinal diversity gradient does is to emphasize very strongly the inextricable link between ecological and evolutionary processes and the short-sightedness of tackling ecological questions in ways that don't admit the existence of an evolutionary dimension to the questions concerned. The current ecological theatre, the players and the nature of their interactions are both the product of evolution in the past and the raw material for evolution in the present. And at present, one of the most widespread sources of evolutionary pressure on a whole range of species is modification of the ecosphere by man. Let's now turn from patterns – both ecological and evolutionary – in the balance of undisturbed nature to the influence of man upon those patterns. In what sorts of ways are we altering ecological stability, and do we represent a threat to its very existence? These are the questions to which the next two chapters are devoted.

15

ECOLOGICAL ENGINEERING

Long before the world went green – meaning the burgeoning of awareness of 'the environment' rather than the evolutionary radiation of land-plants – ecologists were using their knowledge to solve environmental problems. Two major types of problem that have long been recognized, and that are in a sense opposites of each other, are 'overabundance' and 'underabundance' of particular populations. The first of these refers to the unwanted existence, in large numbers, of organisms that are in some way harmful, and therefore embraces the area of pest control; while the second refers to the undesirably low numbers of animals or plants of endangered species, and thus takes on board the field of conservation. In both cases, the ecologist's concern is to manipulate numbers, though of course the changes sought in relation to the two problems are in opposite directions.

Although both pest control and conservation involve the perception and attempted rectification of environmental problems, their pre-dating of the green movement somehow prevents them (especially the former) being seen as green issues – though what exactly the green label includes and excludes is a very subjective matter. So I've decided to treat them separately under the heading of ecological engineering, leaving more topical green issues, like the 'ozone hole', to the following chapter. But I should stress that the distinction between ecological engineering and green issues is a hazy and imperfect one – a point which should become gradually more obvious as we proceed.

Throughout this chapter, you'll see points of contact between the objectives of the 'engineer' and the general ecological processes and principles that have been the subject of the last eight or nine chapters. These include, for pest control: evolution of insecticide resistance, predator specificity and population regulation; and for conservation: minimum viable population size, diversity and extinction. The multiplicity of such points of contact serves to remind us, if a reminder were needed, of the necessity of a firm theoretical basis for our attempts to alter ecological systems. Rush-in-and-do-it procedures, like the casual addition of a new species in the hope that it will control a pest, can, as we'll see, lead to further problems rather than to a solution.

Ousting the Gatecrashers

It's an unfortunate fact that a very large proportion of the surface of the earth is now covered not by natural ecosystems but by their agricultural equivalents. The latter are less diverse, less stable (as we've seen), and less aesthetically pleasing than the natural systems they have replaced, but they are, in a sense, a necessary evil. Clearly, we have to feed ourselves, and with the global human population heading rapidly for six billion and showing no sign of equilibrating, there's no choice but to plan for a future in which agricultural systems will continue to dominate the terrestrial part of the ecosphere.

Given this fact, we can at least act in such a way as to maximize the efficiency of agricultural systems in terms of the amount of produce that they yield to man. Increased efficiency means either more food from an equally agriculturalized world, or the same amount of food from a less agriculturalized world, either of which is desirable – the first because of its potential to relieve starvation, the second because of its potential to release some agricultural land back to a natural state.

Consideration of agricultural efficiency rapidly brings us to insects. Of the total species count of all kinds of organisms, insects make up something like three-quarters, rendering them by far the most prolific taxonomic group and a major ecological force to be reckoned with. And it has been estimated that, worldwide, we lose something like half of our would-be agricultural produce to insects. If we could reduce this fraction to a negligible one, we could in theory sustain present food levels while releasing half of the land area currently devoted to agriculture. A laudable intention indeed, but how do we go about achieving it?

There are two fundamentally different methods of attempting to reduce ('control') the numbers of a pest population, whether it's an insect or something else. These methods are called chemical control and biological control, and I'll deal with them in that order.

The concept behind chemical control is simple. You find a substance that's harmful to crop pests but not to the crop, and you spray it all over your fields, forest plantations, orchards or whatever, at a sufficient concentration to kill the pests concerned. If necessary, you re-spray at regular intervals until the crop is harvested, and then you re-plant and the cycle starts anew.

At least three major problems beset this simple-minded strategy. First, if often doesn't work. One reason for this is that the pest population has evolved resistance to the insecticide concerned, and is therefore unaffected – or only slightly affected – by its application. Another reason is that predators may be more adversely affected than the pests, which can lead to a decrease in predator-induced pest mortality negating the increase in mortality caused by the pesticide. Second, even if chemical control does work in eliminating the pest, it's a short-term solution. I've already mentioned the possible need for re-spraying, which would arise, for example, if new immigration occurs after local pesticide levels have decayed. Finally, most insecticides and other pesticides are injurious to human health and to the health of all sorts of other organisms that they contact – for example those in rivers into which the chemicals are eventually drained by rain-induced 'run off' from fields.

The only sensible reason for using a technique that's inefficient and unhealthy and produces at best a short-term effect is that there's no alternative. Is this the case here? Increasingly, the answer is 'no'. Over the last half-century or more, ecologists have developed an alternative to control by chemicals, namely control by natural enemies – or biological control as it has come to be called. The principle behind this is a little more subtle than that behind the profligate spraying of chemicals, so let's look at it in more detail.

As we've seen in earlier chapters, natural populations usually exist in a state of constrained fluctuation, wherein they hover around an equilibrium value that in most cases is determined largely by their resources or their natural enemies – predators, parasites and the like. In the case of enemy-mediated equilibria, the resources of the population concerned will be found in abundance, while food-limited populations tend to cause their resources to be damaged or scarce.

Agricultural pests are at best like food-limited populations in nature. But they may be even worse than that, because they may never reach a local equilibrium at all but rather have a nomadic existence moving from patch to patch of suitable habitat, decimating each patch as they go. Locusts provide the most extreme example of this latter strategy, but many other pest species behave in a broadly similar, if not quite so dramatic, way.

The principle behind biological control is to introduce into an agricultural system where a pest outbreak is occurring a known enemy of the pest concerned. If the enemy interacts with the pest in a density-dependent manner, then it may be able to regulate the pest population at a low level, rendering it an insubstantial threat to the crop-plant on which it feeds. This strategy is most likely to succeed where the pest is an 'introduced' one from foreign parts. Such pests are commonplace, and one reason why they often flourish is that their natural enemies have been left behind while the locally-available predators aren't adapted to dealing with them. In such cases, the best strategy for the ecological engineer is to seek out natural enemies in the country or continent from which the pest originally came, and to use them as the biological control agent.

A graphical picture of a successful biological control programme of this kind is given in Figure 38. Notice in particular that it's neither intended nor expected that biological control causes complete elimination of the pest. But this is a strength of the method rather than a weakness. Since the aim is to have a long-term solution to the problem, it makes sense to sustain a

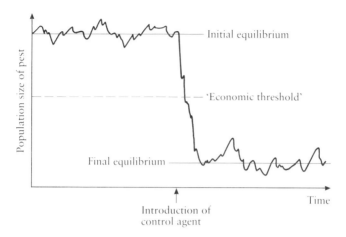

FIGURE 38   An idealized biological control programme.

small enemy population that can rapidly expand to control subsequent pest outbreaks rather than to have the whole pre-dator–prey system 'implode', with extinctions of both pest and enemy.

If we can achieve the sort of result shown in Figure 38, we are clearly out-doing chemical control in all sorts of ways. There is no risk to human health. There is no need to keep re-applying the control agent, as it's self-sustaining. And while pest populations can evolve in response to natural enemies just as they can evolve in response to pesticides, the natural enemy can coevolve, and so 'track' the evolution of the pest, which is something that no chemical can achieve. Furthermore, not only is a successful biological control programme better than its chemical equivalent, it's also likely to be considerably cheaper. So it would seem to have everything going for it.

So much for the principles and the ideal outcome – what results have actual attempts to conduct biological control programmes produced? The answer is just about every conceivable result from total success to total failure, and this inconsistency is probably the biggest single factor contributing to the continuing use of chemical methods.

At least we now understand some of the reasons for past

failures, and can act to ensure that they aren't repeated. One of the most embarrassing sorts of result – which has occurred on several occasions – is that the control agent eats the pest to extinction but then, rather than going extinct itself, begins to consume the crop plant that it was intended to protect. That is, the control agent becomes the pest.

In such cases, the problem lay in the use of a rather generalist predator – or a polyphagous or broad-niched one if you prefer. it therefore makes sense in future to use enemies that are much 'fussier' with regard to their diet. In the case of insect pests, there's a ready-made solution to the problem of predator polyphagy in a group of organisms called the insect parasitoids. These are insects which feed on other insects in a highly specific way that makes them incapable of turning their attentions to crops should their insect prey become unavailable. The parasitoid's mode of operation is like this: the adult female pierces the skin of an egg, larva or pupa of the prey species and deposits an egg inside. Her egg hatches into a parasitoid larva which gradually grows, eating the prey from within. Eventually, what bursts out of the prey's pupal case is not an adult insect of the prey species, but an adult parasitoid ready to go and parasitize further prey. Clearly, this rather hideous lifestyle makes parasitoids ideal biological control agents.

That's all very well, but does biological control by parasitoids actually work? The answer, at least in some cases, is 'yes'. One example involves the winter moth which we met earlier when looking at the general principles of predator–prey interactions. The studies we examined then were done in Britain, which is part of the natural range of the winter moth, since it's a native of Europe. But like many other creatures, accidental inter-Realm transport by man has given it the opportunity to expand into new regions. In particular, the winter moth became established in Nova Scotia, and began to damage forests there on a very wide scale. Following recognition of this problem, parasitoids were introduced from Europe, and two in particular (*Cyzenis albicans* and *Agropyron flaveolatum*) were successful in reducing winter moth numbers. A study carried out before and after introduction of these natural enemies showed that percentage mortality of pupae increased from about 37 per cent to more than 80 per cent following parasitoid introduction, and virtually all of this increase was due – directly or indirectly – to attack by the two main control agents.

It has to be admitted that not all attempts at biological control using parasitoids have been successful, and the reasons why some attempts still fail are not well understood. So we'll have to be patient and continue to research this highly desirable form of pest control if it's to be consistent enough to displace chemical control, with its unwanted side-effects and high cost. Recent innovations include using microbial parasites, and the combined use of several control agents, which is given the somewhat pompous title of 'integrated pest management'. One thing is for sure: the lack of *any* control measures may work all right on some small-scale organic farms, but it won't feed the world.

Nurturing the Endangered

Biological control is an odd sort of endeavour for an ecologist to be involved in, since it turns him into a kind of executioner. Most ecologists *like* animals and plants, including those that the layman finds slimy, dirty or distasteful. So, while there are sound reasons for undertaking the control of so-called 'pests', it isn't a role that comes easily to the average ecologist. Conservation, however, fits like a glove. Here, we are squarely on the side of nature, rather than acting against it. However, partly because of this, it's all too easy to drift into a general conservational ethos of 'doing good to nature', without thinking too carefully about precisely what we are trying to conserve, and why. Let's be a little more careful here.

There are at least four quite separate things that ecologists are interested in conserving, and while they sometimes go together they don't always do so. They are: species, natural biomes, diversity and wilderness. In linking conservation with the problem of underabundance, as I did earlier, I was thinking primarily of the first of these four – i.e. the problem of endangered species, such as those listed in the 'red data books'. But that's really only one branch of the business.

When Charles Elton marvelled at the patchwork of habitats in

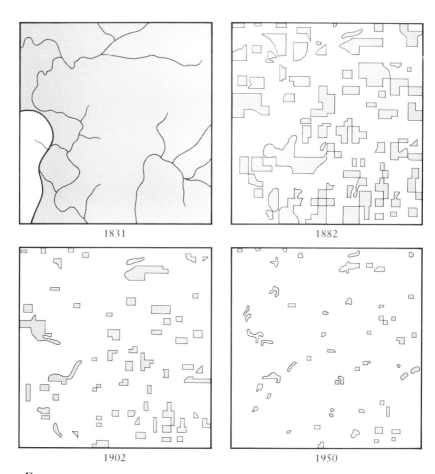

FIGURE 39 Destruction of natural forest by man over a period of 120 years. The region shown is a 10 km square area of Wisconsin, USA.

the Herefordshire countryside, and the need to conserve it, diversity was his central concern.[45] Those of us who are appalled at the disappearance of moorland habitat in northern Britain under an ever-increasing tide of conifer plantations are primarily interested in the retention of our country's last real wilderness. Concern about the disappearance of mixed temperate forest in the United States (see Figure 39) relates to the loss of a natural biome. The closeness of certain whale species to extinction brings us back to endangered species, which is where we started.

Most of these examples have been relatively easy to classify in terms of our four central themes, because they are examples of where the different themes *don't* go together. Loss of a species of whale is a tiny dint in diversity and can hardly be said to contribute a threat to wilderness. And whales certainly don't form a biome. Elton's habitat diversity (which also entails a reasonable degree of species diversity) is almost the opposite of a natural biome, since the patchwork of existing habitats is a result of *destruction* of the naturally occurring climax (temperate deciduous forest) from most of the area concerned. Few would describe Herefordshire as a wilderness, and it's not exactly high on the list of places harbouring the remnants of endangered species. Moorland Britain is regarded by most as a wilderness area, but it's low in diversity, is of dubious 'naturalness' and is also low on species which could be described as endangered. Forest destruction in Wisconsin is not particularly important from a wilderness or endangered-species viewpoint, and the system concerned is only moderately species-diverse.

While such examples serve to indicate the quasi-independence of the four central themes of conservation, destruction of tropical rainforests provides the classic example of their being bound up together. As we destroy this magnificent biome, we destroy with it countless species, many of which have not even yet been discovered. We also destroy the highest species diversity on earth, and what is genuinely one of the ecosphere's 'wildest' places. Here, more than anywhere, we need to conserve for *all* reasons.

So far, we've concentrated on the question of what ought to be conserved. But now it's time to move on to ask how the job can be done. There are three phases to achieving any piece of conservation work. First, we need to *identify* the species, biomes and wilderness areas that are most at risk. Then we have to produce *plans* of how best to conserve them, taking into account such factors as the size, shape and location of nature reserves. Finally, we move to the hardest part, *implementation* of our plans, which is where we come up against the thorny problem of conflicting interests for the use of any particular piece of land.

In general, the first two phases are reasonably well advanced.

I've already given plenty of examples in which the identification of species or systems for conservation is no longer in doubt. Much work has also been done on the planning phase. Among the conclusions to have come out of work on the design of nature reserves are the following. Reserves should be as large as possible, given constraints of funding and of competing demands on the landscape from other needs (housing, industry and agriculture in particular). For a given overall area, reserves should be either continuous or broken up into a few large patches. Fragmentation into many individually small reserves risks violating the minimum viable population sizes for their inhabitants. Creation of 'stepping-stone' reserves or 'corridors' in between far-flung main ones may assist with migration, thus opening the possibility of asynchrony of population processes between localities ensuring overall survival of species despite local extinctions.

So where do we stand? To a large extent we know what we want to conserve and how best to go about it, but can we put it into practice? That is, can we implement our plans? This third and final phase of the conservation process is the most difficult, not least because it's an 'interdisciplinary' venture, involving cooperation between ecologists, politicians and all sorts of land users rather than being something that ecologists can quietly get on with themselves.

It's easy to point to apparent success stories in the implementation of conservational plans. There are, for example, the American National Parks; and in Britain there are also such parks, though much scaled down in size, along with areas designated as AONBs (areas of outstanding natural beauty), SSSIs (sites of special scientific interest) and 'green belts'. But do such designations actually achieve anything, or are they just bureaucratic pieces of terminology?

The main problem is that when possible land uses conflict, conservation tends to get a low priority in the pecking order. This is why green belts get built on and factories are allowed to go on polluting the environment, even in the heart of national parks. The only way that this will change is through a change in public attitudes to the importance of our natural environment. Of course, there are signs that such a change is now beginning to take place, with the recent political successes of the Greens in Europe. But will it last? That question forms the essence of the problem, since transitory shifts towards environmental awareness that disappear a decade on will hardly be sufficient to achieve our ecological goals.

We're clearly drifting, now, towards a more general discussion of green issues, but before we get there (in the following chapter) there's one final aspect of the implementation of plans for conservation that I want to cover. This is the question of how we can achieve such implementation when it concerns areas that are outside our own political boundaries. I'm thinking in particular of the rainforests, the three main remaining areas of which are in Brazil, the Congo and Indonesia. The environmentally aware European or American public can fairly directly influence legislative processes in their own countries, but they don't have much clout in the Brazilian parliament.

I'd like to suggest a fairly radical solution to this problem. There's no point simply continuing to shout loudly and wave our fists at the Brazilians and others concerned. Not only is this futile, but it's actually rather hypocritical. In Britain, for example, about 99 per cent of the natural forest system has been destroyed; and Figure 39 shows that, while the process is somewhat less advanced in the United States, it's going along the same tracks. But if shouting and publicity won't do the job, what will? The answer, of course, is money.

Now I'm not so naïve as to suggest that vast amounts of aid should be handed over to the countries in which areas of rainforest are found, along with some vague advice that the forests be turned into National Parks. No, what I'm suggesting is that we *buy* the rainforests. If an affluent Brazilian can own a London flat or a Californian beach-house, why should not an equally affluent 'Westerner' – or Northerner, to be more precise – own a few acres of rainforest? And there's no need to restrict such sales to individuals: environmentally aware institutions such as Greenpeace could become forest owners also. Indeed, the big lending banks, which don't like writing off the debts owed to them by countries like Brazil but know very well that the chances of ever being fully repaid are remote, could settle for land instead and get themselves some good publicity in the process.

All this will only work if the sales of rainforest have built into them a cast-iron clause stating that no development of any kind should take place. Ownership could be associated with the sole right to non-destructive visits at an agreed frequency, so there would be a 'holiday' incentive as well as the more altruistic one of saving the forest. And that incentive to the new owners would be linked with an incentive to the host governments, because all those extra visits would bring in extra foreign cash.

While rainforest conservation poses major problems because the forests lie outside Western political boundaries, it is at least true that any particular area of forest lies within a *single* country. Conservation of marine systems presents an even more difficult problem because most of the oceans are outside *any* political boundaries. This means that not just one foreign government but many are involved. The well-publicized difficulties in getting an international agreement to conserve whales illustrate just how hard a job marine conservation can be.

One possible solution would be to make more use, in the marine sector, of the 'nature reserve' approach that has been reasonably successful for terrestrial and freshwater habitats. Perhaps many governments could be more easily persuaded to abandon whale-hunting (for example) in specified reserve areas than in general; and the adverse publicity attendant on violating such reserves could be much more dramatic, I think, than that deriving from violation of general whaling restrictions. But to be effective, for such large and mobile creatures, it's clear that reserves would have to be substantial in area.

This treatment of conservation has had what might be called a progressive flavour. We've gone in logical sequence from what we want to conserve (species, biomes, diversity and wilderness) to identifying precisely which species etc. need to be conserved, to the planning and ultimately the implementation of conservation measures. But I want to end by jumping right back to before the beginning of this sequence. I want to pose the question: why conserve at all?

If you ask this question of a large number of ecologists and environmentalists, you'll get many different answers. The motivation behind conservation is both selfish and altruistic, both quantitative and intuitive. Some people want to conserve rain-forests because if we don't there might be drastic climatic changes which would be detrimental to mankind. Others are driven by a gut feeling that it is fundamentally wrong to take entire species and communities that are the product of millions of years of evolution, and wipe them out forever just to gain another few square miles of boring farmland. Some ecologists will argue for an area of mountain and moorland to be conserved because it contains a particular rare species of plant. Others will simply say that the landscape is awe-inspiring, and that standing on a mountain-top you can see for miles, with hardly a human artefact visible from horizon to horizon – an increasingly rare possibility in heavily populated countries like Britain.

In the end, what matters is that those of us who wish to conserve the world's ecological heritage act in concert to maximum effect, regardless of the diversity of our motivation. If we can do so, we may make headway against the entrenched forces of ecological squanderers who until now have largely had the upper hand. If we can't, then the future for rainforest, moorland, whales and every-thing else looks very bleak indeed.

16

SELECTIVE GREENERY

One of the aims of this chapter is to persuade you that your local shopping centre poses a greater threat to ecological balance than does a massive oil spill, such as that from the *Exxon Valdez*. I hope that, at the outset, this seems a bizarre proposition to make. If it doesn't, then I'm wasting my time trying to persuade you, but then again it might be interesting to see whether our similar conclusions are based on similar or different reasoning.

I've used 'greenery' in the chapter title here because the topics with which I'm now going to deal are all perceived as being firmly within the popular image of 'green' or 'environmental' issues, however ill-defined that mental bucket may be. But while all the issues dealt with here will be green, I will certainly not deal with all green issues. The reason is simply that it's better to treat a few topical environmental problems in depth than to mimic a collection of newspaper articles and deal with lots of them superficially.

The potential advantages of selectivity, in terms of depth, carry with them a danger, namely the risk of selecting for discussion issues which are not the most important ones. So how do we decide which *are* the most important? A good starting point, I think, is a hierarchical classification of environmental problems, coupled with a clear statement of precisely what each entails. This should help to put things in perspective, and in particular should enable us to avoid confusing one environmental issue with another.

I saw a classic example of such confusion in a television debating programme. The subject of the greenhouse effect came up, and a leading politician, on being asked about it, started ranting about acid rain. It was left to another of the panellists – a union leader – to point out that the two were not the same. That leading politicians should be so environmentally ignorant is staggering – and worrying too. Let's now take the necessary steps to ensure that we don't follow their example.

Classified Information

One way of classifying environmental problems is presented in Figure 40. The primary division, here, is between problems arising from the removal of things ('resources') from our environment, and problems arising from the addition of other things ('pollutants') to it. Of course, we can't pretend that these two halves of the problem have nothing to do with each other. After all, many of our pollutants arise as side-effects from our use of resources.

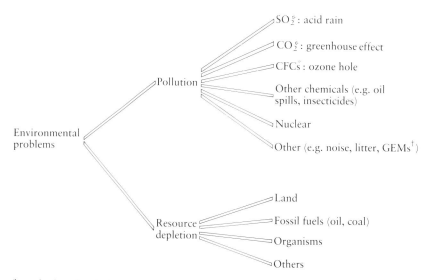

* and other chemicals

† genetically engineered micro-organisms

FIGURE 40   A classification of environmental problems arising out of human activities in the ecosphere.

Nevertheless, the problems attendant on resource depletion and pollution are quite distinct. Reduced mobility caused by the death of the automobile following the final elimination of fossil fuel reserves will hardly, when it happens, resemble the inhalation of petrochemical smog.

Having dealt with the major division of environmental problems, I'd like, now, to go through the finer divisions listed in Figure 40, one at a time, giving a brief, but hopefully clear and useful, description of each. I won't at this stage pass comment on the relative importance of the various topics on our list – that will come later.

We start, then, with acid rain. The origin and nature of this problem are reasonably well understood. The burning of fossil fuels in industrial complexes produces, among other pollutants, sulphur dioxide and various oxides of nitrogen. These are pumped out into the atmosphere, in large quantities, through factory chimneys. They react with water vapour in the air, forming sulphuric and nitric acids, and these destructive chemicals (along with others that may be associated with them) are then carried back down to earth in precipitation, largely in the form of rain, but also in hail and snow. The acid falls on the uppermost layer of plants (trees in a forest, grass in a grassland), and causes a variety of physiological problems, some of which are transmitted, via the food web, to animals. Because of the global distribution of industrial areas and the direction of prevailing winds, most problems associated with acid rain are concentrated in Eastern North America and in Europe.

Next is the greenhouse effect, also known as global warming. This is associated with a different product of the burning of fossil fuels, namely carbon dioxide, and a major source of this is your car (in conjunction with my car, his car, her car and so on a few million times over). The build-up of carbon dioxide in the atmosphere has an effect similar to that of night-time cloud cover following a sunny day. That is, it acts like a greenhouse and lets radiant heat in but restricts the loss of 'degraded' heat. So the more CO_2 in the atmosphere, the hotter it gets. This is a world-wide problem, rather than one that is concentrated, like acid rain, in the countries that are maximally industrial and those that are unfortunate enough to be downwind of them.

That brings us to the hole in the ozone layer. It's a natural feature of the present-day world (but not of its earlier self) that there's a layer of ozone gas high up in the stratosphere. This absorbs a considerable amount of the sun's output of ultra-violet radiation, which as a result never reaches lower down into the ecosphere. However, in the last few decades, several kinds of pollutants that attack the ozone layer have steadily built up in concentration. Notable among these are the 'chlorofluorocarbons' (or CFCs for short), which are produced from a wide range of domestic sources, including some aerosol sprays. CFCs break down, in the atmosphere, into 'active chlorine', which reacts with, and consequently depletes, the ozone. So the ozone layer gets thinner, and, where it was already thinnest (over the Antarctic), it now disappears for some of the year, leaving a hole. Both thinning and 'piercing' of the ozone layer lead to increased incidence, on the ground, of potentially damaging ultra-violet radiation.

The next category is a sort of miscellaneous 'dumping ground' because it takes in all other kinds of chemical pollution. (There are far too many to deal with individually here.) Included in our chemical miscellanea are insecticides, which we briefly discussed in the previous chapter, and oil pollution of the seas and shores resulting from shipping accidents. Since these and other sources of chemical pollution in our dumping-ground category are very heterogeneous, there's really no way we can generalize about their effects. Suffice it to say that chemical pollution is widespread in all major ecological systems, and its effects range from negligible to lethal.

We now move from chemical to nuclear pollution. The difference is that in a non-nuclear chemical reaction (such as the destruction of ozone by active chlorine or the metabolic interference to an insect caused by DDT) the atomic nuclei of the molecules involved in the reaction are relatively stable. To a large extent, ordinary chemical reactions take the form of a shuffling about of peripheral electrons. However, in nuclear reactions, such as those occurring in the sun (fusion) or in a nuclear power station (fission), the nuclei themselves become unstable. In the process, small amounts of matter are converted into massive amounts of energy. We can envisage several sources of nuclear pollution: accidents (at both civil and military installations), dumping of nuclear wastes (largely from power stations), atomic weapons testing (now, thankfully, much reduced), and intentional use of these weapons during wars. Arguably, the hole in the ozone layer could also be said to be a source of nuclear pollution, since it lets in additional radiation from the sun's nuclear 'reactor'.

And now to the ultimate dumping-ground category: all other pollution. That is, pollutants which are not chemical in the usual sense of the word or nuclear. Noise and litter spring to mind as obvious examples. While I'm quite aware that a discarded plastic bottle is 'chemical' in a broad sense, as a pollutant what we're concerned about is its physical and visual intrusion into our environment rather than its chemical reactivity. Indeed, such objects are extremely inert and unreactive chemically, which constitutes their main threat: they may take hundreds of years to break down. They certainly don't fall into the category of bio-degradable, though of course many other kinds of litter, such as newspapers, do.

A recent entrant into the 'other pollution' category is the genetically engineered micro-organism, or GEM for short – bacteria (usually) whose genetic material has been altered by, for example, the artificial insertion of genes from other organisms. The next couple of decades will probably see a large increase in the frequency of field trials involving the release of GEMs into agricultural systems with a view to improving their efficiency in various ways. It's difficult at this stage to predict the extent to which this is likely to cause ecological problems. My guess would be that the hazards of GEMs will turn out to be *less* than anticipated (in contrast to many other forms of pollution), largely because they will be less fit than their 'wild' counterparts and will be eliminated by natural selection. But it would be unwise to rely on this, and careful monitoring of all programmes of GEM-release will be essential.

Having dealt with the various forms of pollution, let's now turn our attention to the matter of resource depletion. First on the agenda is land. Our most obvious using-up of previously ecological land is that associated with urbanization and industrialization.

The sprawls of Greater London, Los Angeles or the Ruhr industrial complex each replace hundreds of square miles of what had been natural biomes. These massive urban-industrial areas are not of course without their own ecology. But the ecology of man, sparrows, bacteria, gardens and parks is a far cry from that of the temperate deciduous forest that once stood on the site of the Tower of London.

Although the 'concrete jungle' is our most visually dramatic using-up of land that could support natural communities, agricultural land use needs to be noted too. The massive grain belt of the central United States is hardly any more natural than Los Angeles, and may well have a lower species diversity. On the positive side, agricultural systems represent a more readily reversible destruction of nature, but on the negative, they cover a vastly greater area than all our urban sprawls combined.

And so to fossil fuels. These fuels (principally oil and coal) are used up in a much more permanent way than land. If we let London crumble, it might be back to deciduous forest in a thousand years. But once we've extracted and used all our oil reserves, then that's it – gone for good. These resources are non-renewable in the most absolute sense of that phrase. We may have become a little blasé about the eventual demise of the fossil fuels and all that goes with them (planes, trains, cars etc.) because of an earlier scare, in the 1960s and early 1970s, which turned out to be premature. I remember well listening, as an ecology undergraduate, to predictions of the end of fossil fuel reserves by 1990. Since then, vast extra reserves have been discovered, thus rendering the predictions obsolete. But by 2090, new reserves may be somewhat thin on the ground, and in relation to environmental issues we need to take a long-term view.

That brings us to organisms. Their inclusion under 'resources', I should stress, is simply a device to keep Figure 40 reasonably small, neat and user-friendly. I hardly think we should view all the plant, animal and microbial populations of the ecosphere simply as cannon-fodder for the machine of human 'civilization'. In some cases, our using-up of organisms takes the form of causing the extinction of species that have no resource value whatever. But that doesn't mean that we should ignore their demise – indeed the field of conservation, as we've seen, is geared to protecting such endangered species. In other cases, species that have become much rarer through human activities do also have value as resources in addition to whatever intrinsic 'value' a species may be said to have in itself. Overfishing provides a spectacular example. But at least organisms constitute a renewable resource, while species do not. So the creation of rarity, while undesirable, is not as final as the causing of an extinction.

I've bundled all other resource-depletion problems into a final miscellaneous category. Again, like the 'miscellaneous pollution' category, there's not much you can say about this one except that it's exceedingly diverse, and of course that the depletion of non-renewable resources, like some metals, carries more long-term consequences than the depletion of renewable ones, such as wood.

Identifying Ecological Threats

Often, when some new environmental problem arises or is first recognized, the newspapers are full of phrases like 'ecological crisis' or 'environmental catastrophe'. Indeed, they've been used so often now, and so inappropriately, that such phrases are beginning to lose their earlier sense of urgency. But we should allow ourselves to be influenced neither by 'hype' nor by the cynicism that tends to follow it. Rather, we should think very carefully about what really does constitute a severe threat to our environment, about whether there are fundamentally different kinds of threat, and about whether any of the problems described in the foregoing section pose sufficient of a threat to put them into the 'crisis' category.

I'd like, first of all, to distinguish between three different kinds of environmental threats. These are Ecological Threats (where the whole ecosystem is at risk), Environmental Health Threats (where mankind is at risk) and Wildlife Threats (where the risk is to particular animals and plants). I'll now give some examples to illustrate the differences more clearly, and then concentrate on the first of the three, which is most appropriate to a book dedicated to the 'green machine'. As we'll see, however, some threats are compound, involving two, or even all, of the three above types.

Let's start with the rupturing of the *Exxon Valdez*. At the time I'm writing this, a vast area of Alaskan coastline is polluted by oil arising from this accident. Thousands of seabirds, mammals, and smaller less newsworthy creatures have died horrible deaths in a black slick which is entirely alien to them. Clearly, this constitutes a massive Wildlife Threat to the area concerned. That, however, is the end of it. Because the incident was geographically restricted to a region of extremely low human population density, and because the terrestrial environment was relatively unaffected, it cannot be said to constitute an Environmental Health Threat. And I wouldn't call it an Ecological Threat either, though that view may require a more lengthy justification.

What's needed to qualify as an Ecological Threat is *the power to destroy an ecological system beyond its inherent resilience* – beyond, if you like, what I called in an earlier chapter the self-levelling capacity of natural communities, including secondary succession. Using this criterion, rather few events qualify as Ecological Threats. Certainly, an oil spill the size of that which came from the *Valdez* does not. Rather, it's comparable to the natural destruction, by a lightning-induced fire, of a large area of coniferous forest. In both cases, the problem concerned (both of them Wildlife Threats), will disappear, due to the natural resilience of the ecological systems involved, over a relatively short timespan. Indeed, to an 'order of magnitude' level of accuracy, the *Valdez* oilspill may be ecologically invisible in not much more than ten years, while a large-scale forest fire may not be completely compensated for until nearer a hundred years after the event.

Let's move on then to an Environmental Health Threat. Global warming is a good example. If, due to atmospheric build-up of carbon dioxide, global temperatures rise by a few degrees, it's possible that much of the polar ice-caps will melt. If they do, then many of our cities, which are concentrated on coastlines and estuaries, may be submerged by the resultant rise in sea-level. This could cause enormous disease problems, not to mention the sheer social disruption, particularly in countries that are low-lying, heavily populated, and poverty-stricken. Indeed, the floods in

Bangladesh have provided a 'model' – albeit thankfully a transient one – of the much more permanent effects that global warming could have.

So far, I've identified events that can be primarily thought of as Wildlife Threats and Environmental Health Threats. But what of Ecological Threats? With such a stringent requirement as destruction of the powers of natural resilience, do *any* of the problems listed in Figure 40 qualify? The answer is a clear 'yes' in two cases: our usurpation of the land, and the threat of large-scale nuclear pollution. Concern about these two in particular constitutes my 'selective greenery' and it is to these issues that I'll devote most of the remainder of the chapter. But I'll return, towards the end of the chapter, to the question of whether there are other Ecological Threats besides these two.

No Room for Nature

In the British Isles, the percentage of the total land-area that plays host to completely natural ecological communities is perilously close to zero. In continental Europe, things aren't much better, though Scandinavia rates considerably higher than the rest. In North America, large areas of natural forest remain, but as we saw in Figure 39, there has been much destruction too. The story is similar elsewhere, from the grain-belt of Siberia to the pathetic low-yield farms that are replacing rainforest in Brazil. Nature is literally being squeezed out of the terrestrial sector of the ecosphere, to be replaced by our farms, industrial complexes and cities. Each year, the global coverage of natural ecosystems goes down, the coverage of artificial systems up. Not only that, but while this process is *potentially* reversible, in practice it tends not to be. So while we could ultimately see the Tower of London revert to forest in a thousand years, my guess is that in a thousand years the Tower will still be there – unless that other Ecological Threat (the nuclear one) envelops us, in which case there probably won't be a forest either.

My opening proposition now makes a little more sense. Your local shopping centre, however large or small it may be, will probably still be taking the place of a natural ecosystem a thousand years hence, when the Alaskan coastline has long since forgotten the *Exxon Valdez*.

Here we come to the crunch. Distinguishing genuine Ecological Threats from what happens to be the newsworthy environmental crisis of the moment has been a useful exercise. We can see the sense in being concerned not so much with events that temporarily destroy ecological equilibria but rather with events that destroy the whole natural means of climbing back to such equilibria. That's fine. But becoming concerned with the *real* issues makes dealing with them doubly difficult. It isn't hard to find volunteers – or even finance, for that matter – to help with a clean-up operation following an oil spill. But you can't seriously advocate the demolition of shopping centres or the abandoning of farms in the interests of allowing nature to regain a foothold on her planet.

The answer, it seems to me, is twofold. First, we should make our urban and agricultural areas as 'efficient' as possible, thus minimizing the land they use. We must be careful, though, to let our intent in this direction be constrained by other considerations. Cities where people are packed into anonymous tower blocks simply won't do; nor will battery-style farming or intensive use of vicious pesticides. But even staying within such constraints, there is much that we can do to improve our land-use over its present, rather chaotic, form.

Such improvements will be of various kinds. Most cities contain a substantial amount of derelict land that could either be built upon, thus reducing infringements of yet more countryside at the city's periphery, or turned into nature reserves, thus improving the quality of life for city-dwellers. Some countries have too much land devoted to agriculture. Making it financially viable for farmers to relinquish some land to nature is then a sound policy – as in the embryonic 'set-aside' practice in the UK. If improvements in biological control and integrated pest management, as discussed in Chapter 15, are eventually marked enough to result in the production of significantly more crops from the same area of land, then set-aside could be carried out on a larger scale. Areas of land reclaimed from industrial usage – such as a spent open-cast mine – could be rehabilitated more extensively, more quickly, and more imaginatively than at present. Thus we might end up with a nature reserve consisting of a mixture of forest, lake and parkland, set in a cluster of small, irregular hills, instead of a large, grassed-over symmetrical mound which looks little more natural than the mine that preceded it.

The second strand in our attempts to let nature regain a foothold should be clear long-term plans relating to the identification of areas that we aim to allow to 'go natural'. This almost brings us back again to the business of designing nature reserves; except that the most desirable project of all – the relinquishing of certain *large* areas of land to nature – is too grandiose in scale to fit the 'nature reserve' label. Not only that, but it won't need constant management, as many small reserves do. Rather, we should be moving towards designating a series of large, internationally recognized 'Nature Zones' which we simply leave alone. We should get away from the idea of nature reserves as things that are perpetually tinkered with and 'managed' according to our rather whimsical views about how nature should behave. I'm not saying that we shouldn't have *some* reserves of the 'manicured' type, but we need some *real* nature too.

Where we should put these Nature Zones will require exhaustive debate, but let me throw in the 'three As' (Alaska, Amazonia and Antarctica) for starters. The politics of implementing such a giant concept will undoubtedly be horrendous (as recent discussions of the future of Antarctica illustrate), but if we really care for natural ecological systems as valuable entities in their own right rather than merely as things to be exploited, then surely we must make the effort.

The Holocaust to Come?

Let's start here by scotching the illusion that the natural world is a nuclear-free zone, and that man's creation of sources of radiation represents something totally new, artificial, and out of keeping with the rest of nature. Quite apart from the fact that the sun is a nuclear fireball, many types of rock are naturally radioactive. These rocks and other natural sources, acting in concert, give rise to what's often referred to as 'background' radiation. There's no escape from this, since it's worldwide, but it does vary somewhat in intensity from place to place.

Background radiation doesn't constitute an Ecological Threat. Indeed, not only has evolution occurred in its presence, but it may have made a contribution to the evolutionary process by providing at least some of the mutational novelties that have led to the evolution of new kinds of organism, including ourselves.

While human-initiated radiation is not, therefore, something entirely new to the ecosphere, it does represent a substantial, and in some cases massive, increase over background levels. In many cases, this poses an Environmental Health Threat. Clusters of leukaemia in the vicinity of nuclear power stations are beginning to look unlike coincidence. And the dumping of radioactive wastes almost certainly involves a tangible health risk too.

However, such sources of radiation, and indeed even nuclear accidents such as Chernobyl, do not by any stretch of the imagination pose an Ecological Threat. The reason for this lies in the vastly different acceptable levels of mortality in the human and natural worlds. If a thousand people are killed by a nuclear accident, that's a disaster of major proportions. But many 'lower' organisms routinely lose that sort of number in a single generation through natural wastage. On the ecological scale of things, the death of a thousand creatures is a drop in the ocean. Put another way, radiation levels of the sort associated with the Chernobyl accident may disturb ecological equilibria (and hence pose a Wildlife Threat both more subtle and more invisible than oil slicks), but they certainly won't destroy the self-levelling powers of natural communities.

Why, then, have I included 'nuclear pollution' as one of my pieces of 'selective greenery' – one of the few environmental problems or potential problems capable of posing an Ecological Threat? The answer is simple: the worst conceivable kind of nuclear pollution – that deriving from a full-scale nuclear war – is capable of destroying the ecosphere as we know it, and it's virtually alone in this respect. At the very moment you are reading this, there are, scattered through the ecosphere, thousands of nuclear weapons primed and ready to be fired. Should they actually *be* fired, whether by intent, by madmen or (perhaps most likely) by computer virus, the worldwide level of radiation would be such as to destroy most life. And even if some were left, as it probably would be, it would take a rather lowly form. The relative sensitivities to radiation of mammals (including ourselves), insects and bacteria are shown in Figure 41. In the hypothetical situation (which I hope will remain that way) where all organisms received a dose of a million rads, there might be nothing left but bacteria. It's a sobering thought that the firing of the world's nuclear arsenal could be achieved in a matter of minutes, and that as a result the ecosphere could be sent backwards by more than a billion years. No other force on earth could even come close to that level of destruction.

Now this scenario, you could argue, is extremely improbable, and I'd have to agree. The probability of such an event occurring today is probably 0.00001 or less. But the uncomfortable fact about probabilities is that in certain circumstances, including this one, they're additive. So if the probability of a nuclear holocaust occurring today really is 0.00001 (for argument's sake), such an event becomes a virtual certainty in 100,000 days – that is in about 280 years' time. And my calculations are, of course, very rough and ready since the daily probability is impossible to estimate accurately. If I'm out by a factor of 10 one way, we're all right for a few millennia. If the other way, then twenty-eight years should do the trick.

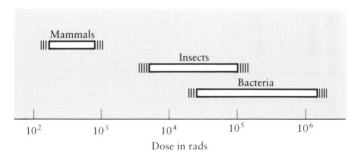

FIGURE 41 The relative sensitivities of mammals, insects and bacteria to varying doses of gamma radiation. (Left ends of bars – the most sensitive species in the groups are affected; right ends – the most resistant species in the groups are affected.)

Given this potential Ecological Threat hanging over our heads, you could be forgiven for wondering why some ecologists spend their time on relatively minor, localized problems like acid rain. In fact, it's perfectly reasonable for them to do so, because we don't simply drop all minor and middle-sized problems when faced with a major one. But what we could *not* be forgiven for is lack of action to attempt to remove the nuclear Ecological Threat before numerous vanishingly small probabilities add up to oblivion – the ultimate 'last straw' syndrome.

But what can we do? I have only one suggestion to make, which is this. All those of us who are aware of the magnitude of the Ecological Threat posed by nuclear weapons, and the suddenness with which it could be invoked, should be tireless in our attempts to inform and to educate. Many politicians doubtless think that nuclear weapons are not qualitatively different from very large lumps of dynamite. Even if they're aware of the nature of radiation, they may have visions of sitting it out in their privileged bunkers, re-emerging when the worst is over, blissfully ignorant of the fact that some radio-isotopes (Plutonium 239, for example) have half-lives of thousands of years. As in other matters, people are most ready to take action on a problem when they feel personally threatened by it. So our task should be to make those in power in *all* nuclear-armed countries feel the nuclear threat sufficiently tangibly and personally that they become motivated to get involved in serious negotiations to remove it. That way, we can perhaps avoid a return to the green-slime ecosphere of the distant past.

Back to the Three Threats

The strategy I have adopted in this chapter has been to classify the main types of environmental problem that human industrial and agricultural activities have generated, to classify the types of threat to the environment that they variously pose, and to concentrate on the two most severe Ecological Threats, namely depletion of the land resource and wide-scale nuclear pollution. While this strategy has obvious advantages, it also carries two hidden dangers –

namely (a) that the three types of threat may be seen as discrete categories with little or no overlap; and (b) that we end up thinking that only two things qualify as Ecological Threats.

Figure 42 is an attempt to prevent these dangers from becoming realized. Here I have indicated for each type of environmental problem identified in Figure 40 (except 'other' categories) what combination of threats it poses. I'd like to devote a little space to discussing the rationale behind the entries in the Ecological Threat column – except, that is, for the land-depletion and nuclear cases which we've already discussed.

I've tentatively listed acid rain as 'probably' causing an Ecological Threat. The reason for my caution is simply that we have to be careful about inferring causal relationships from statistical associations. Where large areas of forest exhibit canopy deterioration and there is found to be an elevated level of acid in precipitation in the region concerned, then there may indeed be a causal link between the two; but experimentation (now ongoing) is needed before we can be sure. The same can be said of the effects of acid rain on aquatic systems, where increasingly a link between acidification and the deaths of many fish and other aquatic organisms looks likely.

Type of pollution or resource depletion	Wildlife threat?	Environmental health threat?	Ecological threat?
Acid rain	√	(×)	(√)
Greenhouse effect	(×)	√	√
Ozone hole	×	√	×
Nuclear (local accident)	√	√	×
Nuclear (global war)	√	√	√
Land depletion	√	×	√
Fossil fuel depletion	×	√	×
Organism depletion	√	×	(×)

√ = yes   (×) = probably not
× = no   (√) = probably

FIGURE 42   Sources of the 'three threats'.

The greenhouse effect certainly *does* pose an Ecological Threat, though of a rather subtle kind. If, due to ice-cap melting and sea-level rising, terrestrial communities near low-lying coastlines become submerged, this clearly will take them beyond their intrinsic 'self-levelling' capacity and lead to their outright destruction. But accompanying this destruction will be the creation of new aquatic ecosystems. It will be a case of 'new ecosystems for old'. And we come back to the point made earlier, namely that the *primary* classification of the greenhouse effect should undoubtedly be as an Environmental Health Threat.

The same could be said for the ozone hole. Here, the main problem will be an increase in the incidence of skin cancer. As bare-skinned organisms, we will be most at risk. There seems little likelihood of the ozone hole leading to community-wide effects, unless of course it becomes so pronounced that mutation rates are raised generally, and even then the question of whether ecological stabilizing mechanisms would be threatened remains open.

The depletion of fossil fuels (as opposed to any resultant pollution) does not in itself pose an Ecological Threat, especially since most such fuels are located far underground (or undersea) and hence are outside the ecosphere. But the depletion of *organisms* is a less clear-cut case. Instances of overfishing in which low fish population sizes lead to reduction or cessation of fishing followed by fish-stock recovery are clearly not taking the systems concerned beyond their powers of resilience. In contrast, depletion of tropical rainforest trees *does* pose an Ecological Threat because the whole forest community depends upon them, and their destruction is usually a one-way process. But because this destruction can be seen as a case of depletion of the natural land resource by agriculturalization, I have taken the view that 'organism depletion' refers more to the overfishing type of scenario, which is why I have labelled it tentatively as not constituting an Ecological Threat.

I should stress that my concentration on Ecological Threats should not be taken to imply that the other two types of threat are unimportant. Clearly, they are. The greenhouse effect, in particular, is capable of posing a very severe Environmental Health Threat, and one that we must invest considerable effort in attempting to predict and prevent. Rather, my concentration on threats to the existence of whole ecosystems simply reflects the overall thrust of the book – towards understanding of, and respect for, the 'green machine' that surrounds us.

Beneath our perception of the diversity of threats covered in this chapter lie two very different emotions. On the one hand, concern for the ecosphere as a human life-support system; and on the other, an interest in the continued existence in their own right of totally natural systems, with their inbuilt balance, in at least *some* of the ecosphere's many regions. That is, we are back to the point at which the book started – the contrast between wonder and anxiety, the two emotions that the balance of nature, and our potential for destroying it, engender. In this chapter, anxiety has been allowed to get the upper hand. But I will end on a positive note, by devoting the final chapter to wonder.

17

UNIVERSAL ECOLOGY

Let's end where we began – in the vastness of space. Our mission, as you'll recall, was to undertake a thousand-planet sampling programme to discover what the typical planetary environment was like. And we saw that the probable outcome of our mission was a realization that most planets were barren and lifeless, and that the prevailing flavour of the universe, from an ecological perspective, was therefore one of 'staticness'. But suppose that, somewhere in the great galaxy of Andromeda, on our homeward stretch and our thousandth and final planet, we hit gold and discovered our first extra-terrestrial ecosphere. What form would it take?

While this question clearly can't be answered, it's interesting to consider the possibilities; and having conscientiously probed the 'facts' about the balance of nature on earth, I think we've earned the right to indulge in a little speculation about the form that balance may take elsewhere.

This is where the difference between ecology and natural history becomes most clearly exposed. It's a safe bet that, whatever form our extra-terrestrial ecosphere takes, there's not a single species there that's also found here. There will be no *Homo sapiens*, no *Drosophila melanogaster* and no *Gorilla gorilla*. So the natural history of Xenobia, as we'll call it, will be quite different from that of earth. All the species-specific details of feeding, nesting and mating behaviour, or whatever is appropriate to the beasts concerned, will be different.

That's not to say, of course, that there will be no furry

mammals, no birds, insects or plants. One thing that we're entirely ignorant of is the extent to which evolutionary processes operating on different planets will follow broadly similar paths. We do know enough to assert that they won't be identical – hence my point about the lack of any species-level equivalences. But if they're different, where will the two evolutionary processes begin to diverge, and how different will their products be?

One possibility that I'd rate very highly is that they'd be different right from the start. But even if that were true, there are still many forms that the difference might take.

We know, from chemical analysis of meteorites and Martian rocks, that the elements comprising alien pieces of matter are the same as those from which terrestrial materials – from rocks to rockets – are made. So the life-forms of Xenobia must be based on a chemistry which is within the bounds of our understanding. But whether that chemistry is carbon-based is another matter. All terrestrial life is based on this element: all the substances of which our bodies are composed – principally, proteins, nucleic acids, carbohydrates and fats – are carbon-based molecules. And the same applies to other animals as well as to all microbes and plants. Perhaps on Xenobia a carbon-based chemistry again holds sway, but it's also possible that a whole alternative molecular basis for living systems has been stumbled upon.

While a silicon-based biochemistry (for example) represents one extreme possibility, the other extreme – in the sense of degree of similarity to terrestrial systems – is that not only are the life-forms of Xenobia based on carbon but that the broad nature of the molecules concerned is similar. That is, they're a mixture of molecular types which we would broadly recognize as nucleic acids, proteins and so on. Even if this were the case, there's still scope for important differences. We can, for instance, anticipate a different genetic code. Studies of some primitive terrestrial organisms and cellular organelles reveal a degree of flexibility of the code that, in the beginnings of molecular genetics, was not appreciated. Perhaps in early biochemical evolution there's competition between groups of organisms characterized by different codes, and there's a tendency for all but one code to be lost or to survive only as a rarity. If so, then it's clearly possible that in extraterrestrial carbon-based ecospheres a different code from our own could become stabilized during early evolution, and persist indefinitely afterwards in all the resultant life-forms.

We've now considered enough scenarios to see the full range of possibilities for life on Xenobia or anywhere else: from non-carbon-based to carbon-based with different molecules, to similar molecules with different interactions (like the code), to similar molecular interactions but a different array of organisms. Given such a wide range of alternatives between which we can't conclusively choose (until we actually make the journey), how can we speculate about the possible ecology of Xenobia – about patterns in its balance of nature if you like?

The surprising thing is that we can do so very easily, because the ecological principles that we've seen in earlier chapters may well transcend the chemistry of the organisms whose ecology we're considering. That is, not only are extra-terrestrial ecospheres likely to be characterized by the same levels of ecological organization that we find on earth – populations, guilds, food webs, communities and ecosystems – but the general ecological *principles* that we've established (and are still establishing) on earth may also apply. Foremost among these principles are density dependence, frequency dependence, limiting similarity, stabilization of prey communities by predator-switching and, more generally, the self-organizing and self-levelling behaviour of successional and climax systems respectively. These principles – the fruits of ecologists' endeavours – may apply on Xenobia and beyond just as do the laws of physics. If so, then ecology is not a 'local' science involving the piecemeal study of one tiny corner of the universe. Rather, it's a science of very wide applicability, and the earth is merely one of many manifestations of its principles.

Of course, we can't expect too much. Other ecospheres, if they exist, may not exhibit patterns in their balance of nature that are directly comparable to our own. Shorter food chains may prevail, latitudinal diversity gradients may be absent and populations may be less patchily distributed, for example. But the *mechanics* of nature's balance – the forces underlying the self-organization and 'homeostasis' of ecological systems – could well be the same regardless. I'll hazard a guess that they are. Such a guess could even be raised to the scientific respectability of a hypothesis, since it's potentially testable, albeit the timelag from proposal to testing may be longer than usual.

Now all this assumes that we're dealing with 'pre-cerebral' ecospheres. That is, those that have yet to suffer the disruption to their fabric that the evolution of intelligent organisms involves. But what of those in which such a beast *has* made its appearance? It seems likely that gross exploitation and alteration of the natural systems will usually accompany the coming to global power of sentient beings, but is it just a passing phase? Is there a post-industrial, post-nuclear, post-Ecological Threat phase in ecospheric histories, and if so what form does it take? We've finally reached the kind of question that it's not even easy to speculate about, so the end of our journey is fast approaching.

Yet there's a final problem we have to address. If, in our early explorations of space, we stumble upon a pre-cerebral ecosphere, we can study it to our heart's content and, I hope, conserve it. Such a discovery would be 'positive' in every respect. But what of the discovery of an ecosphere dominated by beings of our own degree of mental development or beyond? Would this also be a 'good thing' for mankind? I think a little caution is appropriate here.

The Spanish ecologist Ramón Margalef, in his fascinating book *Perspectives in Ecological Theory*,[46] makes the point that one feature of ecological systems is that more advanced or organized systems tend to absorb more primitive ones. This process may also apply to human societies: clearly, contact with Western civilization was bad news for cultures like those of the American Indian, the Eskimo and the Aborigine. If we make contact with beings whose culture is more advanced than ours, we may find ourselves similarly losing our identity. So while such interplanetary contact could be exciting, it could also be dangerous. Curiosity could indeed kill the cat.

Now this shouldn't be taken as a suggestion that we in some way curtail our search for new ecospheres. All exploration has its risks, but our response to recognizing them shouldn't be to curl up into a ball and do nothing. Rather, we should proceed with caution, as realists entering the unknown and not as naïve optimists seeing only the bright side of the cosmic coin.

We've now completed our ecological journey. We've looked in depth at the mechanics of individual ecosystems, taken walks across the earth to compare one system with another, and embarked on imaginary travels in space to enquire about the likelihood and nature of ecospheres other than our own. Having done all that, what are we left with? Most of all, we should be left with a respect for nature, and in particular a respect for those four strands of nature between which we distinguished when examining the aims of conservation: species (from our own to the lowliest insect), natural biomes (such as the rainforests), diversity (in all its senses), and wilderness (which is as undefinable as it is recognizable when you find it). As ecologists, we should aim to further our understanding of these and other facets of the balance of nature, to conduct our lives in such a way that we minimize our destruction of them, and to persuade others to do likewise. Should we fail, the penalties may be great. Should we succeed, the world will be a better place in which to live, not only for ourselves, but also for all our co-voyagers, with whom we spin through a sea of space and time, from an inscrutable beginning to an imponderable end.

GLOSSARY

ABIOTIC   Refers to the non-living components of eco-
                   systems such as air, water and rock.

ACID RAIN   A form of airborne pollution in which acids
                   formed by interaction between industrial
                   waste gases (e.g. sulphur dioxide) and water
                   vapour precipitate back to earth.

ALLOPATRIC   Living in different places. Can be applied to
                   species, to different races of a single species,
                   and to the speciation process through which
                   such races become species in their own right.

AMENSALISM   Ecological relationship between two species
                   wherein the first has an inhibitory effect on
                   the second, but the second has no detectable
                   effect on the first.

AUTOTROPH   Organism which makes its own food rather
                   than consuming other organisms. The main
                   group is the photo-autotrophs (plants); a
                   much smaller group is the chemo-autotrophs
                   (certain microbes).

BIOMASS   The total amount of organic matter present at
                   a particular time and place. Can be applied to
                   a single species or used more broadly.

BIOME   A community type. Examples include tropical
                   rainforest, temperate grassland and arctic
                   tundra.

BIOTA   The complete flora and fauna of a given
                   region.

BIOTIC — Refers to the living component of ecosystems (the community or biota).

BOREAL FOREST — The extensive natural coniferous forest stretching across high-latitude regions of Canada, Scandinavia and Siberia.

BOTTLENECK — Populations are said to have gone through a bottleneck if they have endured a period in which their numbers were severely depleted.

CANOPY — The uppermost layer of foliage in a forest ecosystem.

CHAOS — An erratic pattern of fluctuation in a variable such as population size which is not caused by random variation in extrinsic environmental factors such as the climate.

CLIMAX — The community to which a successional sequence leads. Each area has a characteristic type of climax (e.g. deciduous forest in moist temperate lowlands).

CLONE — A group of genetically-identical individuals produced by asexual reproduction of a single parent organism.

CLOSED — Refers to populations and communities to/ from which there is no, or negligible, migration.

COEVOLUTION — Joint evolution of ecologically interacting species (e.g. predator and prey) in response to their interaction.

COMMENSALISM — Ecological relationship between two species wherein the first has a beneficial effect on the second but the second has no detectable effect on the first.

COMMUNITY — The biotic component of an ecosystem, that is, the total biota of a given area. Sometimes also used for a part of the total biota – bird community, plant community, etc.

COMPETITION — Interaction between individuals or species that is detrimental to both or all concerned. This may occur through sharing a limited resource (exploitation) or through direct antagonism of some kind (interference).

COMPLEXITY The complexity of a system is usually measured as the number of component parts or types of parts. In ecosystems, species richness is a central strand in complexity.

CONGENERS Two or more species belonging to the same genus (adj. congeneric).

CONNECTANCE The number of links in an ecological web as a proportion of the number of possible links. Usually used of food webs, but may also be applied to interaction webs.

CONSERVATION The business of trying to protect endangered species and natural biomes from extinction/destruction by man.

CONSPECIFIC Belonging to the same species.

CONTRAMENSALISM Interaction between two species where the first has a beneficial effect on the second, while the second has a detrimental effect on the first.

CONTROL (a) Attempts by man to reduce the population sizes of species of pests or weeds (not to be confused with regulation); (b) A part of some experimental designs: the 'left-alone' treatment with which other treatments are contrasted.

DENSITY The number of individuals per unit area, volume or resource.

DENSITY DEPENDENCE Phrase used to describe birth rate declining and/or death rate increasing as the density of a population rises.

DETRITOVORE Organism that consumes non-living organic matter (detritus).

DISTURBANCE 'One-off' environmental influence on an ecosystem, such as a fire, flood or landslide.

DIVERSITY In ecology, usually used to refer to species diversity, in which case it is either a synonym

	for species richness or an abundance-weighted measure of richness.
ECOLOGICAL EQUIVALENTS	Species occupying similar positions in different ecological systems. These may be products of convergent evolution.
ECOLOGY	Usually defined as the study of the relationship between organisms and their environment. Redefined here (p. 33) as the study of patterns in the balance of nature.
ECOSPHERE	That part of the earth and its atmosphere in which living organisms are found, extending from a few metres below ground to a few hundred metres above. Synonymous with biosphere. Despite name, more the shape of a shell than a sphere.
ECOSYSTEM	The totality of biota and abiotic substances present in a particular place (e.g. an island or pond).
ECOTONE	Region of intergradation between two adjacent but different communities (e.g. forest bordering grassland).
EFFICIENCY	There are many measures of ecological efficiency. Used here in the sense of the proportion of incoming energy to a trophic level that is transmitted on to the next level up.
ENVIRONMENT	A very general term used to refer to everything outside, but impinging upon, a particular reference point – usually an organism. (Confusingly, physiologists sometimes refer to an organism's internal environment too.)
EQUILIBRIUM	A value which a variable like population size or species richness will stay put at. In the case of stable equilibria, displacements away from that value also lead to returns to it.
EXPLOITATION	A form of competition based on joint use of a limited resource.
FACILITATION	The aiding, in some way, of one species by

	another (with the reciprocal effect, if any, undefined).
FAUNA	The total animal complement of a given region.
FITNESS	A compound of the survival probability and average offspring number of one genetic type in a population relative to that of others. Fitness differences equate with natural selection.
FLORA	The total plant complement of a given region.
FOOD CHAIN	Linear sequence of feeding interactions (e.g. plant/herbivore/carnivore/top carnivore).
FOOD WEB	Net-like pattern of feeding interactions, reflecting the complexity of nature more realistically than 'food chain'.
FREQUENCY DEPENDENCE	Phrase used to describe situations in which the growth rate of a particular species population decreases as its relative frequency in the community (or part of it) increases.
GENUS	The next level of taxonomic grouping above the species. Boundaries more arbitrary than those of species. (Plural = genera.)
GLOBAL WARMING	See Greenhouse effect.
GREEN	Colloquially used to mean environmentally aware/beneficial. Used both as noun ('The Greens') and adjective ('green politics').
GREEN MACHINE	Used here to mean ecosystem, with particular reference to its biotic component, i.e. the community. 'Green' is being used here in its original (colour) sense, to refer to the predominant colour of terrestrial communities.
GREENHOUSE EFFECT	The warming of the earth's climate due to build-up of carbon dioxide (etc.) arising from combustion of fossil fuels. Also known as global warming.
GUILD	A group of species in a particular place utilizing the same resource in a similar way (e.g. seed-eating birds in a terrestrial ecosystem).

HABITAT	Where a particular kind of organism lives, in a fairly general sense. A particular insect's habitat might be temperate deciduous forests, while its micro-habitat (see below) might be the undersides of oak leaves in such forests.
HECTARE	A metric unit of land area. Equivalent to 10,000 square metres or one hundredth of a square kilometre. (Approximately $2\frac{1}{2}$ acres.)
HERMAPHRODITE	An organism which is both male and female. Such individuals (e.g. many snails) may nevertheless cross-fertilize rather than self-fertilize.
HETEROTROPH	Organism that consumes other organisms (alive or dead) rather than making its own food. Animals, fungi and some microbes. Other microbes and plants are autotrophs (see above).
HOLISM	Approach to systems in which they are considered as integrated wholes rather than collections of semi-autonomous parts.
HOMEOSTASIS	Equilibrium-seeking behaviour, usually in the physiological realm. But density dependence can be considered as a kind of ecological homeostatic mechanism.
INTERFACE	The plane of contact between one phase of matter and another (the three phases being solid, liquid and gas). Many organisms (e.g. ourselves) live at interfaces, while others (e.g. fish) do not.
INTERFERENCE	A form of competition based on direct mutual antagonism (e.g. secretion of toxins harmful to the other).
INTERSPECIFIC	Between species. Used as adjective of competition, variation, etc.
INTRASPECIFIC	Within species. Used as adjective of competition, variation, etc.
INVERTEBRATES	Animals without backbones. Very broad category including insects, snails, flatworms, jellyfish, starfish, spiders and worms.

KEY (TAXONOMIC) A way of identifying an organism found in the wild down to (eventually) a recognized species, through a series of nested decisions, based (usually) on morphological characteristics.

LARVA Mobile juvenile lifestage.

LIFESTAGE Recognizable subdivision of life-cycle, such as egg, larva, pupa, adult.

LIMITING SIMILARITY Degree of similarity in resource use beyond which species competing exploitatively are deemed to be incapable of stable coexistence. Hypothesis rather than established fact.

MACROSCOPIC Opposite of microscopic. Large enough to be seen with the naked eye.

MARK-RECAPTURE Technique of population size estimation involving the marking, releasing and recapturing (at least once) of some of the constituent individuals.

MICRO-HABITAT That part of an overall habitat within which a particular species tends to be found (e.g. under bark, in crevices in rocks, etc.)

MODEL A simplified view of a complex system which, despite its simplicity, is intended to capture the essence of the system concerned. Models may be mathematical, graphical or pictorial, or take the form of laboratory microcosms.

MONOCULTURE Predominance of one species to the virtual exclusion of all others (or at least those of similar broad type). Many agricultural crops are routinely grown as monocultures.

MONOPHAGY Consuming only one species of prey (adj. monophagous).

MORPHOLOGY Study of the body-form of organisms (adj. morphological – relating to body-form).

MULTIVARIATE Having many variables. All ecological systems are of a multivariate nature.

MUTATION Alteration of DNA sequence leading, in some cases, to a new phenotype. The primary source of genetic variation which acts as the raw material for evolution.

MUTUALISM   Ecological interaction beneficial to both (or all) participants. Pollination of plants by insects is an example.

NATURAL SELECTION   Increase in frequency of those genetic variants in a population which have the greatest fitness under the prevailing environmental conditions.

NICHE   What organisms do in natural communities, with particular reference to what they eat (animals) or the conditions they require in order to flourish (plants).

ORGANELLE   A distinct sub-unit within a cell, in the same way that organs are distinct sub-units within an organism. Each organelle has a particular specialized task. For example, the nucleus houses the cell's genetic material.

ORGANISM   Individual. For example 'a deer', 'a fly' and so on. Hard to apply to some species or groups of species (e.g. grasses) because of vegetative reproduction and the existence of clones.

OPEN   Refers to populations and communities, but used in two senses: (a) migration occurring to/from neighbouring communities; (b) non-wooded (of terrestrial habitats/communities).

OZONE   A gas (chemically related to oxygen) which forms a layer in the atmosphere and absorbs some of the sun's ultraviolet radiation.

PARASITOID   An insect which parasitizes other species of insects. Usually the adult female parasitoid lays an egg inside a juvenile (e.g. larval) stage of the host, which is then consumed from within by the parasitoid larva when it hatches.

PERSISTENCE   The degree to which an ecological system endures over time.

PERTURBATION   Alteration in the state of an ecological system.

	Often used in the sense of perturbation by an experimenter to see how the system responds to certain changes.
P<small>EST</small>	Subjective description of an unwanted organism (usually animal), e.g. insect consuming crop-plant in agricultural ecosystem.
P<small>HENOTYPE</small>	All aspects of an organism except its DNA (genotype). Usually, specific phenotypic characters are referred to – e.g. body size, pigmentation pattern.
P<small>HOTOSYNTHESIS</small>	The manufacturing, by plants and some microbes, of organic compounds from carbon dioxide, water and sunlight.
P<small>LANKTON</small>	The masses of tiny creatures that float at or near the surface of bodies of water, both fresh and marine. Includes both plants (phytoplankton) and animals (zooplankton).
P<small>LATES</small>	The great slabs of the earth's crust that are in constant, albeit very slow, motion with respect to each other.
P<small>OLLUTION</small>	A rather general term for substances and forms of energy that are present in an environment and have detrimental effects on it. Usually restricted to the products of human activities.
P<small>OLYMORPHISM</small>	The existence, within a population, of two or more discretely different genetic variants.
P<small>OLYPHAGY</small>	Consuming many (taxonomically diverse) species of prey (adj. polyphagous).
P<small>OPULATION</small>	All the individuals of a particular species living in a particular place.
P<small>REDATION</small>	The consumption of part or all of a live or recently killed organism by another. May be used in a broad sense to include parasitism and herbivory, or a narrow sense to refer solely to animals eating other animals.
P<small>RODUCTION</small>	The laying down of organic material as a

result of photosynthesis (primary production) or consumption of other organisms (secondary production).

PROPAGULE   A group of individual organisms migrating from an already-populated area to a new one, and establishing a population there. May consist of anything from wandering animals to wind-blown seeds.

PROTOZOAN   Small unicellular 'animal' such as *Paramecium*.

PUPA   The stage in the life-cycle of some insects between larva and adult. A stage with much internal, developmental activity but no external, ecological activity.

PYRAMID   Name given to a stack of trophic levels in which each is less substantial than the one beneath it.

RANGE   The total geographical distribution of a species within the ecosphere.

REALMS   (Wallace's Realms, so called after the nineteenth-century British naturalist, Alfred Russel Wallace; also known as biogeographical realms, regions or provinces.) The six major divisions of the terrestrial sector of the ecosphere, whose floras and faunas differ considerably from each other. Correspond very roughly with continents.

REDUCTIONISM   Approach to systems in which they are broken down into parts for the purpose of studying them.

REFUGE   An area of relative safety from predators.

REGIONAL POOL   All species available for the colonization of a particular island, patch of habitat, etc. of which those that do colonize represent a subset.

REGULATION   A population is said to be regulated if density-dependent forces tend to bring it back to-

	wards an equilibrium size (or band of sizes) following natural or experimental perturbations.
RESOURCE	Something that a population needs, and that it can potentially use up (e.g. food, water, space).
RICHNESS	Usually used in the form of 'species richness', which simply means the total number of species present in a community.
ROBUSTNESS	The degree to which a system's stability properties are independent of its precise construction.
SAMPLE	A small but representative part of something much bigger, used to facilitate study.
SIBLING SPECIES	Very closely related species that are virtually indistinguishable on morphological grounds. Sibling species are almost always congeneric, while many congeneric pairs or groups will not be siblings.
SPECIATION	The splitting of an ancestral species into two or more daughter species, based on the evolution of reproductive incompatibility between some of its constituent populations.
SPECIES	A group of organisms that are at least potentially able to reproduce among themselves.
STABILITY	The tendency to return to an equilibrium following perturbation away from it.
STASIS	Staying still at a particular value. Not to be confused with stability.
STROMATOLITE	A matted structure produced by some blue-green algae (= cyanobacteria). The earliest known fossils are stromatolites (3.5 billion years ago); but living stromatolites are still found today in some parts of the world.
SUBSTRATE	The surface upon which organisms live. Examples include soil, bark, leaves and skin.
SUB-WEB	Part of a food web or interaction web, usually

defined in terms of its culmination in a specified top predator.

SUCCESSION The sequence of communities that takes place between an area's colonization and its eventual biotic stabilization in the climax community.

SWITCHING A behaviour pattern exhibited by some predators in which they are capable of altering their diet so that they concentrate on the species of prey currently most abundant.

SYMPATRIC Living in the same place. Usually applied to different species.

SYSTEM A collection of interacting parts producing definite patterns of behaviour. Organisms, populations and (of course) ecosystems are all examples.

TAXON (Plural = taxa.) A group of related organisms. Taxa are arranged hierarchically, with the smaller ones (e.g. species) included in the larger ones (e.g. classes). The study of taxa and their interrelationships is called taxonomy.

TOXIN A poisonous substance, sometimes produced by one type of organism as protection against others.

TRANSECT A linear array of sampling sites.

TROPHIC Feeding. Most common usages are 'trophic levels' (plants, herbivores, carnivores) and 'trophic interactions' (e.g. predation).

UNGREEN An invented adjective that I use to describe 'pure' ecology – the study of ecological systems in their natural state, in contrast to the study of problems arising from man's impact upon them (green ecology). Ungreen ecology is in a sense pre-green (like pre-clinical medicine) and is emphatically not anti-green.

WEB A net-like pattern of ecological interactions.

WEED   Often used only for that subset of interactions involving one species feeding on another (food web); but may be used more comprehensively to include *all* types of ecological relationships (interaction web).

Subjective description of an unwanted organism (usually plant). For example, grass is considered a weed in a flower bed, but not (of course) in a lawn. The corresponding term for animals is 'pest'.

REFERENCES

1   Taylor, R. J. 1984. *Predation* , Chapman & Hall, New York, p. 14.
2   MacArthur, R. H. & Wilson, E. O. 1967. *The Theory of Island Biogeography*, Princeton University Press, Princeton.
3   Elton, C. S. 1958. *The Ecology of Invasions by Animals and Plants*, Methuen, London.
4   Lewontin, R. C. 1974. *The Genetic Basis of Evolutionary Change*, Columbia University Press, New York.
5   MacArthur, R. H. 1972. *Geographical Ecology*, Harper & Row, New York, p. 1.
6   Williams, C. B. 1964. *Patterns in the Balance of Nature*, Academic Press, London.
7   Arthur, W. 1987. *Theories of Life*, Penguin, Harmondsworth, Middlesex.
8   Gause, G. F. 1934. *The Struggle for Existence*, Williams & Wilkins, Baltimore.
9   Begon, M. 1979. *Estimating Animal Abundance*, Arnold, London.
10 Stenning, M. J., Harvey, P. H. & Campbell, B. 1988. Searching for density-dependent regulation in a population of pied flycatchers, *Ficedula hypoleuca* Pallas. *Journal of Animal Ecology* 57: 307–17.
11 Williamson, M. 1972. *The Analysis of Biological Populations*, Arnold, London, p. 30.
12 Strong, D. R. 1986. Density-vague population change. *Trends in Ecology and Evolution* 1: 39–42.
13 Darwin, C. 1859. *On the Origin of Species by Means of Natural Selection*, John Murray, London.
14 Brown, J. H. & Davidson, D. W. 1977. Competition between seed-eating ants and rodents in desert ecosystems. *Science* 196: 880–2.

15 Park, T. 1954. Experimental studies of interspecies competition II. Temperature, humidity and competition in two species of *Tribolium*. *Physiological Zoology 27*: 177–238.

16 Elton, C. S. 1927. *Animal Ecology*, Sidgwick & Jackson, London.

17 Arthur, W. 1987. *The Niche in Competition and Evolution*, Wiley, Chichester.

18 Butlin, R. K., Collins, P. M. & Day, T. H. 1984. The effect of larval density on an inversion polymorphism in the seaweed fly *Coelopa frigida*. *Heredity 52*: 415–23.

19 MacArthur, R. H. 1958. Population ecology of some warblers of north eastern coniferous forests. *Ecology 39*: 599–619.

20 May, R. M. & MacArthur, R. H. 1972. Niche overlap as a function of environmental variability. *Proceedings of the National Academy of Sciences of the United States of America 69*: 1109–13.

21 Maynard Smith, J. 1974. *Models in Ecology*, Cambridge University Press, Cambridge, p. 67.

22 Lawton, J. H. 1984. Non-competitive populations, non-convergent communities and vacant niches: the herbivores of bracken. In *Ecological Communities*, ed. D. R. Strong, D. Simberloff, L. G. Abele & A. B. Thistle. Princeton University Press, Princeton.

23 Hassell, M. P. 1976. *The Dynamics of Competition and Predation*, Arnold, London.

24 Pimm, S. L. 1984. The complexity and stability of ecosystems. *Nature 307*: 321–6.

25 Lawton 1984 (see ref. 22), p. 68.

26 Hairston, N. G., Smith, F. E. & Slobodkin, L. B. 1960. Community structure, population control and competition. *American Naturalist 94*: 421–5.

27 Pimm 1984 (see ref. 24).

28 Pimm, S. L. & Lawton, J. H. 1977. Number of trophic levels in ecological communities. *Nature 268*: 329–31.

29 Williams 1964 (see ref. 6).

30 Paine, R. T. 1966. Food web complexity and species diversity. *American Naturalist 100*: 65–75.

31 Odum, E. P. 1971. *Fundamentals of Ecology*, third edition, Saunders, Philadelphia.

32 ibid.

33 Arthur, W. & Mitchell, P. 1989. A revised scheme for the classification of population interactions. *Oikos 56*: 141–3.

34 Connell, J. H. & Slatyer, R. O. 1977. Mechanisms of succession in natural communities and their role in community stability and

organization. *American Naturalist 111*: 1119–44.

35 Clements, F. E. 1936. Nature and structure of the climax. *Journal of Ecology 24*: 252–84.

36 Hutchinson, G. E. 1965. *The Ecological Theater and the Evolutionary Play*, Yale University Press, New Haven.

37 Glaessner, M. F. 1984. *The Dawn of Animal Life*, Cambridge University Press, Cambridge.

38 Stanley, S. M. 1987. *Extinction*, Scientific American Books, New York.

39 Van Valen, L. 1973. A new evolutionary law. *Evolutionary Theory 1*: 1–30.

40 Dawkins, R. 1986. *The Blind Watchmaker*, Longman, London.

41 Wynne Edwards, V. C. 1962. *Animal Dispersion in Relation to Social Behaviour*, Oliver & Boyd, Edinburgh.

42 Rees, W. J., 1965. The aerial dispersal of the mollusca. *Proceedings of the Malacological Society of London 36*: 269–82.

43 MacArthur 1972 (see ref. 5).

44 May and MacArthur 1972 (see ref. 20).

45 Elton 1958 (see ref. 3).

46 Margalef, R. 1968. *Perspectives in Ecological Theory*, Chicago University Press, Chicago.

Suggestions for Further Reading

Anyone wishing to read further in the ecological sphere will find a bewildering variety of books from which to choose. I list below some which you may find interesting and helpful. They are grouped into four categories for each of which I indicate the sort of subjects covered and the level of approach.

A. General Texts

Of the many general ecology texts available, three are, in my opinion, particularly good, namely –
Ecology: The Experimental Analysis of Distribution and Abundance, by C. J. Krebs, Harper & Row 1985 (third edition – with a fourth on the way).
Ecology: Individuals, Populations, Communities, by M. Begon, J. L. Harper and C. R. Townsend, Blackwell 1986.
Evolutionary Ecology, by E. R. Pianka, Harper & Row, 1988 (fourth edition).
All of these should be readily understandable to readers who have completed *The Green Machine*. They go into more technical detail than I have done, of course, as befits a text, but they do so in a way which makes clear the relevance of all the examples given to forging ahead with our understanding of ecological systems generally.

B. Population Interactions

A central theme in *The Green Machine* has been the analysis of ecological communities in terms of interactions between their constituent populations. The following books consider such interactions in greater depth.

The level is fairly advanced in all cases.

Predation, by R. J. Taylor, Chapman & Hall, 1984.

The Niche in Competition and Evolution, by W. Arthur, Wiley, 1987.

Herbivory, by M. J. Crawley, Blackwell, 1983.

C. Community Structure

The two books listed below (which are again fairly advanced) provide an interesting contrast. MacArthur's *Geographical Ecology* provides a picture of community structure in which the influence of competition predominates, while Strong et al.'s *Ecological Communities* incorporates a diversity of approaches including the view that the effects of competition on community structure have not yet been demonstrated and may have been exaggerated.

Geographical Ecology: Patterns in the Distribution of Species, by R. H. MacArthur, Harper & Row, 1972.

Ecological Communities: Conceptual Issues and the Evidence, edited by D. R. Strong, D. Simberloff, L. G. Abele and A. B. Thistle, Princeton University Press, 1984.

D. Applications to Practical and Social Matters

The following three books have little in common except that they represent applications (in the broadest sense) of ecology. The first introduces the principles of conservation, the second the principles of biological control, and the third the principles upon which the 'green' approach to politics is based. All three are written at a level appropriate for the general reader.

Biological Conservation, by D. W. Ehrenfeld, Holt, Rinehart & Winston, 1970.

An Introduction to Biological Control, by R. van den Bosch, P. S. Messenger and A. P. Gutierrez, Plenum Press, 1982.

Seeing Green: The Politics of Ecology Explained, by J. Porritt, Blackwell, 1984.

INDEX